SpringerBriefs in Geography

SpringerBriefs in Geography presents concise summaries of cutting-edge research and practical applications across the fields of physical, environmental and human geography. It publishes compact refereed monographs under the editorial supervision of an international advisory board with the aim to publish 8 to 12 weeks after acceptance. Volumes are compact, 50 to 125 pages, with a clear focus. The series covers a range of content from professional to academic such as: timely reports of state-of-the art analytical techniques, bridges between new research results, snapshots of hot and/or emerging topics, elaborated thesis, literature reviews, and in-depth case studies.

The scope of the series spans the entire field of geography, with a view to significantly advance research. The character of the series is international and multidisciplinary and will include research areas such as: GIS/cartography, remote sensing, geographical education, geospatial analysis, techniques and modeling, landscape/regional and urban planning, economic geography, housing and the built environment, and quantitative geography. Volumes in this series may analyze past, present and/or future trends, as well as their determinants and consequences. Both solicited and unsolicited manuscripts are considered for publication in this series.

SpringerBriefs in Geography will be of interest to a wide range of individuals with interests in physical, environmental and human geography as well as for researchers from allied disciplines.

More information about this series at http://www.springer.com/series/10050

Vincenzo Buongiorno

Suburban Retail Spaces

Formative and Transformative Process

 Springer

Vincenzo Buongiorno
LPA Laboratorio di Lettura
e Progetto - DiAP Dipartimento di
Architettura e Progetto
"Sapienza" Università di Roma
Roma, Italy

ISSN 2211-4165   ISSN 2211-4173   (electronic)
SpringerBriefs in Geography
ISBN 978-3-030-54990-9   ISBN 978-3-030-54991-6   (eBook)
https://doi.org/10.1007/978-3-030-54991-6

This Springer imprint is published by the registered company Springer Nature Switzerland AG
The registered company address is: Gewerbestrasse 11, 6330 Cham, Switzerland

Foreword

The book by Vincenzo Buongiorno concerns the formative processes, the contemporary structures and the future perspectives of the retail space, the space dedicated to commerce, sales and exchange. I believe this is an important topic to be investigated in contemporary times.

Relegated to the margins of the city and the attention of architects, retail structures are today one of the few forms of truly vital public space, inheriting, in some way, the characteristics of a tradition of specialized commercial routes typical of the European city, from nineteenth-century shopping streets, to Parisian *passage*, to large department stores formed as real urban knots at the end of the nineteenth century.

With many problems and differences: these structures, widespread in Europe in the second post-war period, especially through the North American version of the big shopping centers, are also, in reality, a place of great contradictions, being public spaces privately owned for sale and profit.

They also subtract vitality to the traditional commercial spaces made up of stores and workshops. Contrary to traditional exchange spaces, strongly rooted in the urban fabric from which they arise, contemporary retail spaces originate at the edge of the city constituting external polarities, often of a territorial nature. They act as distribution tools at the end of production/distribution chains of global dimensions of which they express characters and values: economic pragmatism, the *assemblage* character, a particular disposition to fast change. For this reason, they belong more to international trade networks than to local urban form. An antinomy that can constitute, from my point of view, as a designer, a fertile predisposition to experimentation, a resource.

Current studies on this topic are mostly concentrated on the functional and quantitative aspect of the problem, within a handbook production that tends to neglect the core of the problem, essentially urban. On the other hand, the more properly architectural studies (such as the excellent The Harvard Design School Guide to Shopping, the result of a research directed by Rem Koolhaas) are based on critical-perceptive tools that can hardly propose general design methods.

The work by Buongiorno acknowledges this condition of the state-of-the-art by proposing a point of view, in my opinion, completely innovative. Through a work that constitutes, at the same time, a reading and a project, Buongiorno proposes to read the elements that form the structure of retail spaces not as buildings, but as a unit of an urban aggregate forming fabrics.

The way of reading things, I am convinced, is also a way of designing: the author reads the spaces for trade and exchange as fabrics because he proposes that, as such, they can turn into real parts of the city, bind themselves organically to the territory.

It is, therefore, a critical reading that, as such, cannot be neutral but contains a choice, is oriented to the design of the new as a term of a process (think about the crisis of these structures due to e-commerce) and start of a new phase.

Giuseppe Strappa
Sapienza University of Rome

Acknowledgments

First of all, many thanks to Prof. Giuseppe Strappa, for his generous guide in the research process at the base of this book, for the dialectical, critical dialogue and relevant knowledge transfer. Thanks to the research group of Lpa Laboratory—Lettura e Progetto dell'Architettura, to Prof. Paolo Carlotti, Anna Rita Donatella Amato and Antonio Camporeale, for the rich dialogue about architecture and morphology.

Thanks to Giuliana Buongiorno and Caroline Miranda Feetam, for the fundamental revision of texts.

Finally, important thanks goes to my Family, for the support, all along the way.

Contents

About the Author

Vincenzo Buongiorno 1987, Architect and Ph.D. in architectural and urban morphology and design. Member of the research team in the "LPA—Laboratorio di Lettura e Progetto dell'Architettura" laboratory, DiAP—Dipartimento di Architettura e Progetto, "Sapienza" Università di Roma, Italy. His research activity focuses on the contemporary built reality morphological interpretation, with special interest in specialistic organisms at building/urban/territorial scales, and on the design outputs that can be produced through scientific investigation. Research and teaching experiences carried out in several contexts such as Italy, Canada, Mexico, Portugal and Cyprus. Professional experience in architectural, urban, and landscape design gained as a freelance architect and in the firms: JLCG João Luis Carrilho da Graça Arquitectos and GAP Global Arquitectura Paisagista, Lisbon, Portugal.

Chapter 1
Methodological Introduction, Research Structure and State of the Art

1.1 A Method for Studying Retail Space

The present research aims at interpreting the formative and transformative process of shopping environments and spaces, seen as aggregates/fabrics and investigated by using urban morphology scientific tools.

A topic that is rarely present in architectural scientific literature investigated, on the one hand, mainly through perception and psychology analytical tools, within a metropolis in continuous transformation and abnormal growth made of fragments and random structures. On the other hand, the texts that dealt with the suburban shopping environment's architectural and urban design topic seem mostly confined to a pragmatic and taxonomic handbook function.

Thus, this book constitutes a contribution to fill a gap found in the scientific literature, specifically the lack of a methodical investigation on the suburban shopping environment morphology. The research is carried out using urban morphology methodological tools, and more specifically those belonging to the typological-processual approach (Strappa 2018).

In the adopted methodological perspective, the interpretative investigation of the formative and transformative process, based on the built environment and its historical documentation, is at the same time an act of knowledge, transformation and design (Caniggia 1997; Marconi 2003).

> … reading and design cannot but coincide: reading does not precede the project by providing documentation and information; reading, the critical interpretation of things, is a constituent part of the project itself. It is therefore necessary to consider the reading technique[…] as "the art of the possible": the form not only as a visible aspect of what is, but also of what could be and what could have been (Strappa 2012, 25).

In the research presented in this book, as in the investigated morpho-typological process, the project therefore has a central role as the last (future) phase of the formative and transformative process of territorial/urban/aggregative/building organism.

© The Author(s), under exclusive license to Springer Nature Switzerland AG 2020
V. Buongiorno, *Suburban Retail Spaces*,
SpringerBriefs in Geography,
https://doi.org/10.1007/978-3-030-54991-6_1

As a reconstruction of an operating and generative process, the research proposed here is projected to the design phase and to the opportunities of transformation for an existing built heritage characterized by many critical issues that need a rebalancing to respond to the present and future changing needs and challenges.

The specific research object is the Shopping mall: *Regional and Super regional shopping center* (Mackeever and Griffin 1977, 1–3), interpreted and read as *suburban fabric characterized by a commercial specialization.*

A survey on the structure and formative process of these special environments appears particularly useful nowadays, to understand the numerous questions concerning the consumerist system that underlies the functioning of commercial spaces. The virtualization of many aspects of contemporary life, including the retail through e-commerce phenomenon, produced a strong crisis that threatens the existence of the great malls built in the western world since the second post-war period and generates an urgent need for knowledge about existing structures, for their transformation and for the new structure's design.

The contemporary debate on the urgency of an operational and transformative knowledge (Underhill 2005; Chung et al. 2001; Dunham-Jones and Williamson 2011; Teufel and Zimmermann 2015; Strappa 2017), able to go beyond the pragmatic and empirical approach, is very rich and reflects such a condition that is problematic and critical as well as fertile for new development.

The research takes place in an area dedicated to the interpretation of the urban organism looking at its complexity and contradictions, in its more peripheral expressions.

The urban organism, its edges and component elements such as shopping malls, usually located in a framework that shows them as an indistinct magma of fragments of fortuitous cities, devoid of any possibility of organic relationship between them, are studied instead according to a more general interpretation at several levels that seeks to redefine the component elements and their roles within the building environment's organism in its formative and transformative never-ending process.[1]

Despite being the product of economic forces, which are very strong and often alien to the settlement's cultural area, the suburban mall and the related peripheral context to which it belongs, like any other component of the territory, can be seen and understood as a territorial sub-organism in a becoming process, as ground transformed by man's activity, through the recognition of its productive and settlement potential, identifying "the shape as a visible aspect of a transforming structure ..." (Strappa 2012, 19).

This work therefore intends to propose, even in the contemporary conditions of crisis, an organic interpretation of the territory where contemporary commercial fabrics settle. It is not so dissimilar, within certain limits, from that of the historical territories, which are still "reality, that is the positive and univocal construct

---

[1]Analogous studies and research have been developed on the topic of suburbs and periphery and can be taken as a reference (Strappa 2012; Vachon et al. 2004, Case Sheer and Stanilov 2004);.

of the collaboration between man and nature"[2] (Muratori 1967) and are legible as conformed by fundamental anthropic actions and dynamics, such as that of walking, aimed at connecting and reaching hierarchical polar places.

In the investigation's development, going on with the reconstruction of the formative process of the commercial environments, of their paths that are apparently very different from the traditional paths and their polarizations, we reach a critical situation, that of the contemporary phase, in which, through an abrupt acceleration, a quantitative and qualitative change in the nature of routes, their shape, size, course and traveling capacity is produced. Likewise, we see polarization fragmentation and low-density dispersion.

It is evident that this new condition, considering the great changes and the technical evolutions that have contributed to originate it, continues to respond to fundamental human needs and issues, such as that of movement and settling in a place.

Investigating the formation and transformation of such new and problematic contemporary territories through the tools used in the investigation of historical territories allows us to insert the phenomenon into the great flow of territorial transformations, giving a diachronic and diatopic perspective to the contemporary city, as a typical expression and phenomenon, considered in the current debate as an exceptional element limited only to the current age and, therefore, unique and not comparable.

This book wants to oppose the relative and relational value that every element and phase of the process has within an *organism*, understood as a place of solidarity and coherence even in the apparent fragmentation of reality,[3] (Strappa et al. 2003, 15); contrary to the scientific literature on the suburban city and on *Regional and Super regional shopping centers* dominated by a logic of parts dispersed in the territory according to economic and "artistic" reasons as "absolute values".

The objective of the research is to observe and read a relatively new and unprecedented phenomenon of contemporary anthropic space with its problems and complex dynamics, within the framework of the building environment's formative and transformative logic/chronologic cycles.

---

[2]Muratori (1967), quoted from Cataldi, G. (2015), "Territorio" on wikitecnica: http://www.wikite cnica.com/territorio/—consulted on 06/08/2019.

[3]About the *organism* notion, specifically territorial organism: "[…]Nozione indispensabile alla comprensione dello spazio abitato dall'uomo: attraverso la quale la pura constatazione dei fenomeni frammentati del mondo costruito si organizza in struttura e il particolare, che a prima vista appare isolato e disperso nell'infinita poliedricità del reale, può trovare la sua collocazione all'interno di un pensiero sistematico che tenti di ordinare matrici formative, processi, esiti." "[…] An indispensable notion for understanding the space inhabited by man: through which the pure observation of the fragmented phenomena of built environment can be organized in structure and detail. Thus, each single element which at first sight appears isolated and dispersed in the infinite polyhedricity of reality, can find its place within a systematic thought that tries to order formative matrices, processes, outcomes.".

In this broader and more organic framework, it is possible to trace back the shopping mall's formative and transformative dynamics and to compare it with other typical phenomena of contemporary anthropic space, looking diachronically/diatopically, with their precedents and analogies.

1.2   Research Structure

Within the synthetically defined methodological horizon the research is structured as follows:

This Chapter

A descriptive, methodological and introductory framework on the research topic and on the existing studies concerning it;

Chapter 2

Definition of the formative and transformative process elements. In this chapter, through a reasoned, textual and graphic glossary divided into reading scales (territorial/urban, aggregative, building) and dyadic couples of elements (routes/nodalities, base building/special building), interpretative elements of mall's built environment are introduced that allow to observe it through the lens of the processual typological approach of urban morphology;

Chapter 3

Formative and transformative process interpretation, divided in phases and in various reading scales; the process interpretation is carried out through a comparative analysis of existing or the only documented case studies available in the literature;

Chapter 4

The chapter dedicated to an in-depth study of the problems faced by the suburban mall and to the definition of possible evolutions of the formative and transformative process interpreted and described in the previous chapter, for the transformation of the existing constructions and for new projects;

Chapter 5

Conclusions.

1.3   State of the Art

The conceptual geography of studies on commercial spaces is quite rich and complex. One of the strongest aspects of complexity is given by the interdisciplinary nature of the topic.

Groupings of studies and research can be identified and those testify to the different methods of investigation and different approaches to the topic: pragmatical-manualistic, historiographical, socio-anthropological and typo-morphological.

To find initial investigations on the topic of modern commercial spaces, urban (such as arcades, passages, etc.) or suburban (fairs, markets, and then later on, shopping malls), we must look at architectural and urban design handbooks. In these books there is, on the one hand, a discussion animated by a classificatory and positivist spirit, such as that of the famous manual by N. Pevsner (Pevsner 1986), heir to a long-standing tradition of illuminist origin; on the other hand, other handbooks aim at the reconstruction of a morphological process that shows the mutual relationships among types (base residential and special ones) considering also the mutual exchange on the major scales, urban and territorial. This is seen in the work by the Italian G. Donghi (Donghi 1927), in which it is possible to find a rich and detailed description and analysis of "new" urban special commercial spaces, full of implicit references to their relationship and derivations from other special historical consolidated types (monasteries, palaces, etc.) through a process of transformation and urban knotting (Strappa 2014, 119).

Still in the context of urban commercial spaces located within the traditional urban fabric, and as regards several experiences of historical reconstruction concerning above all the type of the *Commercial Gallery* and the *Arcades* (Mackeith 1986) and to the *Department store* urban type (Longstreth 2010) which then migrated into the suburb, there are some interesting studies aiming at the investigation of *Passage* and *Galerie* morphology as a type, with a diachronic and diatopic approach, such as the studies by J. F. Geist and B. Lemoine (Geist 1995; Lemoine 1989).

Most recent studies, predominantly historical, on multilevel urban commercial environments, underground or at higher levels, complete the landscape of studies on urban commercial spaces (Deglise 2008; Frampton et al. 2012; Yoos and James 2016).

The first systematic investigations that focus, instead, at the suburban territory and at new commercial spaces developed and built there, date back to the studies of U.L.I.—Urban Land Institute and I.C.S.C.—International Council of Shopping Centers. Both associations, founded in the United States between the 1930s and the 1950s, respond with their research and publication activities to the need of bridging the existing knowledge void in relation to a new built environment development mode, that of suburbanization, very quickly diffused in the United States' car-oriented society and territory. These associations, acting as important research centers and reference for the scientific literature specialized in contemporary retail, produced—in the specific case of U.L.I.—a series of handbooks, published in several editions and authored by different scholars throughout the whole twentieth century and in continuity, in the early years of the twenty-first century.[4] In addition to the production of handbooks, many of the advancements of the research on the subject took place

---

[4]From *"The community builders handbook"* published in 1947, seminal publication, conceived to address and rationalize suburban expansion dynamics, to more specific handbooks on shopping center space (Mackeever and Griffin 1977; Casazza 1985; Casazza 1999; Kramer 2008).

in the reports, published annually or six-monthly (I.C.S.C.—*International Council of Shopping Centers*—https://www.icsc.org/news-and-views/research).

If the associations sponsor research responds to the cognitive and applicative interest of the suburban retail world's economic actors, some more independent research and publications by architects professionally interested to this topic constitute important references to have a clear and balanced overview, in which the "spontaneous" tendencies of the models proposed in the associations studies are integrated by "critical" and creative tendencies elaborated by planners, architects and designers.[5]

In the set of publications that share these characteristics (Baker and Funaro 1951; Beddington 1991; Ketchum 1948), those authored by the Austrian-American architect Victor Gruen are particularly relevant. In collaboration with the economist Larry Smith, Gruen wrote the text that will be considered thereafter, the reference for any mall design since the 1960s in America and Europe (Gruen and Smith 1960). At the same time, his editorial work, in parallel and reflecting designed and building works, is a sort of litmus paper of the debate and, more generally, of the relationship among architects and designers with the topic of the suburban mall throughout the twentieth century. The first acclaimed editorial work, written in collaboration with L. Smith, celebrates the construction and mature consolidated type of suburban mall; this first book is followed by other more critical and reflective editorial projects, which tend to analyze the suburban model through a close comparison with the values of the traditional downtown urban fabric (Gruen 1965). This analysis process reaches its peak with the reflections on *pedestrian malls* (Gruen 1973), simultaneously theorized in books and built in reality (e.g.: Multon Pedestrian Mall, Fresno—California), carried out in the 70s decade, fertile years of reflection on the model of suburban and car-oriented development embodied by the shopping mall, on the wave of a more general rethinking after the 1972 oil crisis.

The recent text authored by Coleman (2006), an architect professionally engaged in the design of suburban commercial spaces, constitutes an important reference to a successful attempt at a global and interpretative overview on shopping environments in different contexts and eras.

The collection of essays by Harvard graduate design school students (Chung et al. 2001), coordinated by R. Koolhaas, also belongs to the group of critical and interpretative works. The essays contained in the book, besides specifically investigating the individual sub-topics within the Shopping space general topic, pays great attention to its contemporary problems, such as changes in retail purchase models, due to the introduction of e-commerce and its effects in the construction of commercial physical space at different scales.

Then there is further research which, although situated in other disciplinary fields albeit closely related, constitutes an important reference for the phenomenon's reading and for interpreting the built environment's formative and transformative process.

The historiographical works by Scarpellini (2001, 2006, 2008), which focused on the supermarket's, mall's and large-scale retail trade's diffusion dynamics in the

---

[5]Reference to the notions of spontaneous and critical consciousness: Caniggia (1979), 39–43.

North American and European areas with particular attention to the Italian context, are precious examples. In the socio-anthropological disciplinary field, important references are the studies by P. Underhill (Underhill 2005) on mutual relationships among commercial spaces and purchasing behavior, and those on the relationship between city and consumption dynamics (Baudrillard 2010; Codeluppi 2000, 2014).

Moreover, in the reading of formative processes of commercial specialized route network (passage, galleries, etc.) the historic-philosophical editorial project, unfortunately unfinished, by W. Benjamin on nineteenth century Paris stands as a milestone (Benjamin 1986).

Moreover, there is a further group of studies and research that, going beyond the usual divisions between urban/suburban context and pre-industrial/industrial era, are aimed at the investigation of commercial spaces by looking at their morphological structure, transversely, in a diatopic and diachronic perspective. Within this group there are the texts by Maitland (1985, 1990) in which suburban malls and urban complexes are closely related from the morphological and historical point of view. There are also texts by Longstreth (1997, 2000) and Liebs (1995), which, while taking advantage of a more historiographical than morphological approach, contribute to delineate the evolutionary traces of the suburban mall formative process, starting from the urban context and stressing the mutual influences with the suburban one.

The studies by Moretti (1999, 2004, 1998) focused on a shopping mall type reading with special reference to the North American context (Canada and the US) and European (France) can be situated in this group as rare examples of proper typo-morphological analysis.

Tangentially, some general studies on the suburban city and its transformation stand in this group sharing methodological tools (Case Sheer and Stanilov 2004; Christensen 2008; Dunham-Jones and Williamson 2011; Vachon et al. 2004).

Looking instead toward methodology and to the typo-processual urban morphology approach in the literature, there is a gap in the study of the formative and transformative process of commercial, urban or suburban environments. Except for the aforementioned studies by G. Moretti on the North American and French area, and for some recent publications related laterally to the present research (Strappa and Buongiorno 2019), investigations on the Italian and Mediterranean context are not found. Above all, the literature lacks research on the design and/or transformation process of the existing constructions.

Furthermore, few existing studies linkable to the topic are characterized for being successions of readings and analytical investigations aimed at knowing or identifying typological characters, but always considering the mall "only" as a special building.

There is a lack of references to the relationships between mall and territorial/urban/aggregative organism.

Finally, studies on the mall's internal structure, as a sort of micro-city, at the aggregative and urban scale, are absent, or strongly insufficient.

Some research experiences that look at urban peripheries as integral parts of the urban and territorial organism (Strappa 2012), however, constitute a methodological model for the study of suburban commercial environments for a deep and organic knowledge of its structures.

This book wants to be a contribution to reduce these gaps in the scientific literature (Strappa 2017, 2–5), believing that in order to transform and design the contemporary commercial building environment, it is necessary to deeply understand the problematic structure of these spaces.

The theoretical premise of the analysis proposed here is that the mall is readable as an urban fabric, an urban fabric fundamental both for the (mostly) American suburban sprawling city and for the denser European suburban serial city, as a place of exchange, communication and community building.

While constituting important portions of anthropic space, these fabrics are currently experiencing an important crisis and, at the same time, they are becoming candidates to be pivots of a strategic transformation. The effects of this transformational dynamics go beyond, the large perimeters of commercial space, to innervate the urban and territorial organism.

These fabrics, in their proper and deep nature, are still weakly investigated and can be at the center of a transformation in a broader sense: "from machines for selling to new urban centers" (Gruen 1973), following the formative/transformative process dynamics that they share transversely with all the other anthropic spaces.

References

Baker G, Funaro B (1951) Shopping centres: design and operation. Reinhold, New York

Baudrillard J (2010) La società dei consumi. I suoi miti e le sue strutture. Il Mulino, Bologna

Beddington N (1991) Shopping centres. Retail development, design and management. Butterworth architecture, Oxford

Benjamin W (1986) Parigi, capitale del XIX secolo. I "Passages" di Parigi. Giulio Einaudi editore, Torino

Caniggia G (1997) Ragionamenti di tipologia. Operatività della tipologia processuale in architettura. Alinea editrice, Firenze

Caniggia G, Maffei GL (1979) Composizione architettonica e tipologia edilizia: 1. Lettura dell' edilizia di base. Marsilio, Venezia

Casazza J (1985) Shopping center development handbook. Urban Land Institute, Washington

Casazza J (1999) Shopping center development handbook. Urban Land Institute, Washington

Case Sheer B, Stanilov K (2004) Suburban form: an international perspective. Routledge, New York/London

Christensen J (2008) Big box reuse. The M.I.T. Press, Cambridge-MA

Chung CJ, Inaba J, Koolhaas R, Leong ST (2001) The Harvard design school guide to shopping/Harvard design school project on the City 2. Taschen, Koln

Codeluppi V (2000) Lo spettacolo della merce. I luoghi del commercio dai passages a Disneyworld. Bompiani, Milano

Codeluppi V (2014) Metropoli e luoghi del consumo. Mimesis, Milano

Coleman P (2006) Shopping environnements. Evolution, planning and design. Elsevier architectural press, Oxford

Deglise F (2008) Montréal souterrain. Sous le béton, le mythe. Héliotrope, Montréal

Donghi D (1927) La composizione architettonica. Distribuzione. Sezione 3: Sedi di società commerciali, di aziende giornalistiche, librarie ed editrici, locali e fabbricati per commercio (botteghe, negozi, magazzini di vendita, empori, bazars, gallerie), stabilimenti sanitari, ricoveri, ospizi, asili vari. Utet, Torino

Dunham-Jones E, Williamson J (2011) Retrofitting suburbia. Urban design solutions for redesigning suburbs. John Wiley & sons, Hoboken-NJ

Frampton A, Solomon JD, Wong C (2012) cities without ground: a Hong Kong guidebook. Oro editions, New York/Singapore

Geist JF (1995) Le passage. Un type architectural du XIX siècle. Pierre Mardaga éditeur, Ixelles-Belgio

Gruen V, Smith L (1960) Shopping towns USA: The planning of shopping centers. Reinhold, New York

Gruen V (1965) The heart of our cities. The urban crisis: diagnosis and cure. Thames and Hudson, London

Gruen V (1973) Centers for the urban environment: Survival of the cities. Van Nostrand Reinhold, New York

Ketchum M (1948) Shops and stores. Reinhold, New York

Kramer A (2008) Retail development handbook. Urban Land Institute, Washington

Lemozzine B (1989) Les passages couverts en France. Délégation à l'Action Artistique de la Ville de Paris, Paris

Longstreth R (2010) The American department store transformed. 1920-1960. Yale University Press, New Haven, London

Longstreth R (1997) City center to regional mall. Architecture, the automobile, and retailing in Los Angeles, 1920-1950. The MIT press, Cambridge -MA

Longstreth R (2000) The Drive-In, the supermarket, and the transformation of commercial space in Los Angeles, 1914-1941. The MIT press, Cambridge-MA

Liebs CH (1995) Main street to Miracle mile. American roadside architecture. The John Hopkins University press, Baltimore/London

Mackeever JR, Griffin N (1977) Shopping center development handbook. Urban Land Institute, Washington

Mackeith M (1986) The history and conservation of shopping arcades. Mansell publishing limited, London and New York

Maitland B (1985) Shopping malls: planning and design. Construction press, London

Maitland B (1990) The new architecture of the retail mall. Architecture design and technology press, London

Marconi P (2003) Gianfranco Caniggia, architettura e didattica. In: D'Amato Guerrieri C, Strappa G (eds) Gianfranco Caniggia. Dalla lettura di Como all'interpretazione tipologica della città. Maria Adda Editore, Bari

Moretti G (1998) Analyse morphologique des centres commerciaux régionaux et des tissus urbains qui les contiennent: le cas de l'agglomération de Québec - Thèse de maîtrise. Université Laval, Québec city

Moretti G (1999) Shopping centres and the urban fabric: the evolution of their relationships. In: Corona R, Maffei GL (eds) Transformations of urban form: from interpretations to methodologies in practice. Proceedings of the Sixth International Seminar on Urban Form, Firenze, July 23-July 26-1999. Alinea editrice, Firenze

Moretti G (2004) Entre globalisation et réalités locales -centres commerciaux et formes urbaines à Los Angeles, Montréal et Paris. McGill University, Toronto

Muratori S (1967) quoted by Cataldi G (2015) "Territorio" on wikitecnica. http://www.wikitecnica.com/territorio/-consulted on 06/08/2019

Pevsner N (1986) Storia e caratteri degli edifici. Palombi Editori, Roma

Scarpellini E (2001) Comprare all'americana. Le origini della rivoluzione commerciale in Italia 1945-1971. Il Mulino, Bologna

Scarpellini E (2006) Esselunga. Agli albori del commercio moderno. Art, Bologna

Scarpellini E (2008) L'Italia dei consumi. Dalla Belle Époque al Nuovo Millennio, Editori Laterza, Bari/Roma

Strappa G, Ieva M, Dimatteo MA (2003) La città come organismo. Lettura di Trani alle diverse scale. Mario Adda Editore, Bari

Strappa G (2012) Studi sulla periferia est di Roma. Franco Angeli, Milano

Strappa G (2014) L'architettura come processo. Franco Angeli editore, Milano

Strappa G (2017) Nuovi confini. Territori di ricerca della morfologia urbana. In U + D Urbanform and design, n. 07/08. L'Erma di Bretschneider, Roma

Strappa G (2018) Reading the built environment as a design method. In: Oliveira V (ed). Springer, Cham-Switzerland

Strappa G, Buongiorno V (2019) Commercial Urban fabrics updating. Retail globalization and shopping cultural areas. In: Charalambous N, Zafer Cömert N, Hoşkara Ş (eds) CyNUM 2018 conference Urban Morphology in South-Eastern Mediterranean Cities: challenges and opportunities Proceedings. Cyprus Network of Urban morphology, Nicosia-CY

Teufel P, Zimmermann R (2015) Holistic retail design. Reshaping shopping for the digital era. Frame Publishers, Amsterdam

Underhill P (2005) Call of the Mall: the geography of shopping. Simon and Schuster, New York

Vachon L, Luka N, Lacroix D (2004) Complexity and contradiction in the aging early postwar suburbs of Québec City. In Case Sheer B, Stanilov K (eds) Suburban form: an international perspective. Routledge, New York/London

Yoos J, James V (2016) Parallel cities: the multilevel metropolis. Walker art center, Minneapolis-MN

Chapter 2
Suburban Shopping Mall as an Urban Fabric

It is possible to read the formation and first transformation of territorial structures through a fundamental dyad of opposite and complementary terms composed by 'routes' and 'settlements', linked to motion and stopping according to the primordial needs of man to feed and protect himself, attributing to the term 'settlement' the meaning of temporary or stable structure consisting of a set of dwellings organically related to a complementary productive pertinent area (Strappa et al. 2003, 17).

In order to read a very complex phenomenon, some concepts that proved to be fundamental in the study of historical urban fabrics are used here. Their use, although mediated and updated in relation to a particular study object, is justified by the conviction that even though changing the study object, some expressions and settlement modalities, the anthropic behavior in living and transforming the territory can be traced back to a general nucleus that is common and shared by men of all times and places. This concept is meaningful in the diatopic and diachronic sense.

The description of the process elements is organized in dyadic couples, which are formed by routes and settlements or nodalities,[1] on the one hand, and by base and special elements or building on the other.

The description is also developed according to three different scales: territorial/urban; aggregate/fabric and building/commercial unit.

The identification and description of the elements organized in dyadic couples at different scales as the subject of this chapter, while making specific references to previous phases of the formative/transformative process, relate especially to the current phase, and thus describes today's mall understood as a suburban fabric with commercial specialization.

At the end of each paragraph, a graphic scheme accompanies the textual description of elements.

---

[1] *Nodality*, generated at the intersection of two or more continuous, is considered in the book as the complementary and opposite term of routes. It contains potentially the notion of Polarity, being the latter's results of a specialization and sublimation process of the former.

© The Author(s), under exclusive license to Springer Nature Switzerland AG 2020   11
V. Buongiorno, *Suburban Retail Spaces*,
SpringerBriefs in Geography,
https://doi.org/10.1007/978-3-030-54991-6_2

2.1 Territorial and Urban Scale

As regards to the dyadic couple formed by routes and nodalities, at the territorial/urban scale, two groups of elements can be identified (Fig. 2.4):

– *Routes*, which—seen within an interscalar perspective—constitutes simultaneously territorial structural elements and load-bearing elements for the urban organism; this group includes, in the current contemporary process phase, mostly highly specialized routes (e.g.: highways, expressways, motorways, etc.);
– *Nodalities*, a group that includes settlements, characterized by specialization and functional segregation. Whether they are settlements or aggregates/fabrics characterized by commercial or mixed use specialization, both urban-*in town*, like *pedestrian malls*, or suburban-*out of town*,[2] like *regional shopping centers* or *business edge cities* (Dunham-Jones and Williamson 2011, 172–200), they all share a higher level of specialization in the current mature phase, reached through distinct processes.[3]

2.1.1 Special Commercial Routes

The shape and structure of contemporary territorial routes, referring to the European and North American cultural areas linked to the analyzed case studies, are the result of a remarkable and continuous transformative process of progressive specialization that, in the last century, has undergone a relevant acceleration.

As a result of this process, which started by acting on a mixed-use route—therefore not a specialized kind of route—strictly rooted to the territory from which it arose and which is changed by the route—the specialization process leads to a new kind of route that is the specialized one (Figs. 2.1 and 2.2).

In this book almost exclusively the commercial declination of specialized routes will be analyzed, although the process is rather similar to other types of non-commercial specializations.

The new specialized route responds to the need of "avoiding", through a *bypass route* (Liebs 1995) the vehicular traffic of urban centers, clogged due to the increase of vehicles in circulation. A specialized motor vehicle route such as the American *Urban Highway* that leaving the city, passes through its increasingly undefined perimetrical edge, becomes an axis for the traffic at a territorial and regional level, constitutes an example of this type of route.

The territorial/urban specialized motorway routes' network[4] is characterized for:

---

[2] The division between commercial settlements in town and commercial settlements out of town comes from a vast scientific literature on the subject (Coleman 2006).

[3] In the urban context the specialization process is diachronic, progressive and starts from the base fabric, whereas in the suburban context it is synchronic and generates a fabric already characterized, from its origin, by a high specialization level.

[4] Similar reflections could be made, widening the horizon, even with railways and air transport.

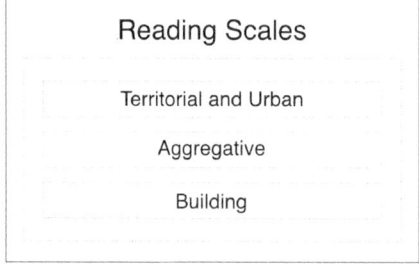
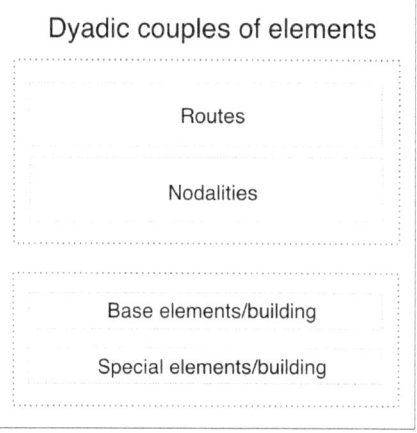

Fig. 2.1 Typomorphological reading's tools *Source* V. Buongiorno

- being made up of *discrete* routes[5] with a limited number of access points, interruption nodal points, such as access/exit nodes/poles for payment toll, or multilevel nodalities connecting different traffic lines (specialized *flyover* nodes, with overpasses and/or subways);
- a strict selection of vehicle types allowed to use the route, made by considering the movement/transport system (e.g.: pedestrian, cycle, road, rail), weight limitations or speed limitations;
- the clear separation of traffic flows in different directions, within the network and each route, in channels dedicated to different traveling speeds and vehicles types. This character explains the great technical attention dedicated to the realization of multilevel nodes (overpasses, underpasses, flyovers) in these networks, designed precisely to allow this traffic flow separation, to make them not only complementary to each other but also independent;
- the impermeability of the route in relation to the territory, the absence of communications between the route and the built-up area on its pertinent strips.[6] This building aggregation, when existing, although rising on the route, is not tied in an organic relationship of solidarity with the route itself, but instead by an insufficient purely visual relationship.

---

[5]Different from the traditional routes network, characterized by a strong continuity of routes, by solidarity between routes and the "surrounding" fabric/territory generated by them. *Discrete* is, therefore, a route that is traced indifferently in its context, characterized by a limited number of specialized access points, and many continuity solutions in its shape and use.

[6]A route's pertinent strip is the envelope which includes the route, the units and the building aggregates developed on it and their related backyard pertinent areas. From the pertinent strips of an aggregate, it is possible to define the hierarchical organization, development phases. Pertinent strip's concept testifies to the generative importance of the route in the formative process of traditional cities and territories (Caniggia and Maffei 1979).

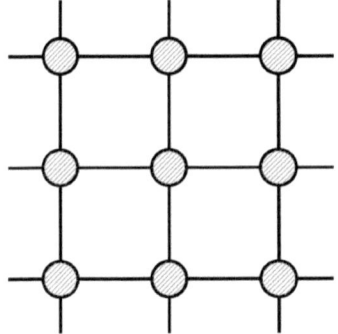

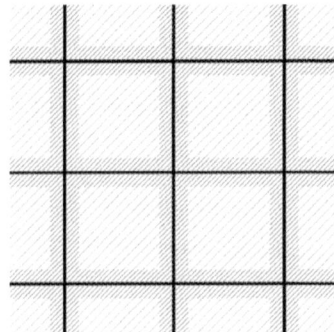

Specialized territorial routes' network
with settlements located in major nodalities within bypassed areas

Traditional territorial routes' network
and settlement on their pertinence stripes

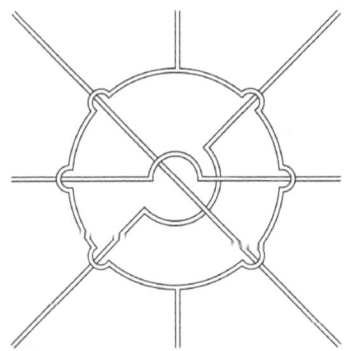

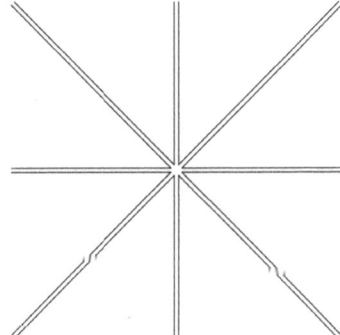

Special multilevel *fly over* nodality

Traditional nodality

Fig. 2.2 Compared territorial routes scheme *Source* V. Buongiorno

Within the networks of specialized routes, it is possible to identify a wide range
of routes, and as the number of route types involved in the network vary, it is also
possible to find:

– the increase or decrease in the level of complexity of the system;
– the variation of its internal hierarchy and organization, over all in its critical points,
  the exchange/access nodes.

Although the consideration of a complex multimodal network is very interesting
and useful for understanding the rich variety of situations existing in a network of
specialized routes (such as the network of specialized routes of a real contemporary
territory, which includes, in addition to the routes, railways, pedestrian and bicycle
paths and exchange nodes with maritime and air routes), this book focuses on regional
and superregional shopping malls considered as commercial specialized suburban
fabrics, and the most attention will be given to the motorway special routes network.

These fabrics with commercial specialization are, in fact, in most cases positioned along the route or next to the motorway system's nodalities, due to their inherent dependence on the car-oriented system. Generally speaking, the motorway route from the famous antecedents of the British *turnpike road* (Hindley 1972, 61) to the first modern highways[7] excludes, due to its specialized character, the fabric generation dynamics on its pertinence strips[8] (in contrast to the typical traditional route that acts as *accentrating axes*), placing itself instead as a *dividing line* (Strappa 1995, 82).

Its linear buffer zones on the two sides of the route (analogous to the traditional routes' *pertinence strips;* Caniggia and Maffei 1979) are the place for technical equipment or, as in the case of American *parkway* roads, can host landscape systems that also perform the function of changing the visual scenario to limit the effect of monotony, which is very dangerous for road safety (Hindley 1972, 129).

There are, however, also situations in which on the specialized motorway route's pertinent strips the formation of fabric can be observed. This condition mainly occurs in periurban and suburban areas, in *commercial strip* settlements (Liebs 1995, 10–35) or in the so-called *miracle miles* (Longstreth 1997, 127–141). In the latter case, on the edges of the traditional city, it is possible to observe first the settlement of special commercial units, and secondly the development of a real fabric, made up also of serial units, on the specialized highway route's pertinent strips.

This commercial fabric forms by following the rules of the spontaneous traditional urban fabric and by setting up the units on the route's pertinent strips, as its emanations. Being developed in adherence to the route, being almost its scenography, it is characterized by a strong intrinsic limit since birth: it shows the contradiction of having interpreted as a *centralizing axis* a special route, the highway, which is instead a *dividing line*. The resolution of this incongruity, which stands at the origin of the clogging problem of these "new" downtowns, as will be deepened in the formative process interpretation (Chap. 3), consisted in the further specialization of routes in the section affected by commercial fabric settlement and related clogging.

This specialization is carried out by implementing further traffic flows divisions: the *bypass*, horizontally or vertically developed (via underground tunnels or flyover overpasses at higher altitudes) allows vehicular traffic to be diverted around or over/under the special fabric, meanwhile specializing both the route and the fabric, through quantitative and qualitative transformation of fabric's internal routes (e.g., by minimizing the car accessibility, or by totally pedestrianizing, or even through the mechanization of the route by equipping it with *tapis roulants* and/or travelator).

The realization of the special bypass route marks a high specialization level for the network of motorway routes and its nodalities/polarities.[9]

---

[7]On the modern specialized automotive routes/highway's formative process, see Chap. 3.

[8]In this new horizon, specialized route network's nodalities increase greatly their importance, if compared with the nodalities of the traditional routes network. In addition to constituting the continuity solutions of the route, making it more discreet, in many cases *Nodalities* are transformed through a spontaneous specialization process, that can eventually transform them into *Polarities*.

[9]The specialization with a bypass generates new nodalities for the network, pushing it to an advanced organic behavior. The bypass ring allows contemporary and differentiated traffic flows with limited

The specialized route observed at the urban scale is characterized by different degrees of divisive influence on fabric (see urban railways, urban highways, avenues, etc.).[10] The dividing character implies that, even when the specialized route[11] is engaged in the urban organism and aggregate, the greatest concentration of urban activities moves "from a centralized traditional fabric" toward the polarities/nodalities of the specialized routes' network.

The logic of the special route and its bypass node is also profitably applied to the urban scale. Through the bypass, aggregates and routes are identified and considered as areas and axes of aggregation, very accessible through the dividing lines' car and tramway traffic but not crossed by them (or equipped with filtered/regulated/reduced crossing).

These new accentrating routes and the specialistic fabric that contribute to innervate form together specialized settlement, which acts as nodalities/polarities for the specialized routes' network (e.g.: project for the "Multon Pedestrian Mall", in Fresno—California, by the Austrian-American architect V. Gruen).

2.1.2 Special Commercial Settlements

The identification of territorial settlement's nodalities and polarities, in a system governed by an organic logic, is strictly connected with the network of routes that structures this territory. The nodalities of the routes' network tend to coincide with the existing settlement aggregation/nodality.

In any case, within the logic of modern design intended as a result of critical awareness, conditions change and become more complex. On the one hand, the specialized route network is designed to better connect and serve existing settlement/territorial nodes, following a diachronic process made of progressive specializations of the network and existing settlements (such as bypass route fabrics for example), on the other hand, routes' network design can be used for catalyzing new settlements' synchronic formation in new planned nodalities.

A first distinction can be identified between:

---

crosses and changes, above all without stops to prevent clogging. At the same time, it distributes and makes the bypassed/specialized fabric very accessible and integrated.

[10]The construction of the rail transport network in North America provided examples of very strong dividing lines, even within the urban perimeter, as in the case of railways allocated on the urban route of the nineteenth-century Manhattan (Rudofsky 1981); as regards Avenues (Hindley 1972, 134).

[11]It is useful to remember that the specialized route in urban context results in many cases from a *restructuring* process (Caniggia and Maffei 1979). It is obtained through demolition within the fabric designed to build a direct connection, quite indifferent from the existing urban fabric, between two critically chosen polarities (e.g.: Paris Haussmann, Corso Rinascimento in Rome).

– settlements formed diachronically, resulting from a "spontaneous" process, progressively specialized through the evolution of the base fabric. An example of this case is the American *downtown* and the *historical centers* in European cities or, even if with a lower involvement of the base fabric in the process of specialization, as well as the primary suburban fabrics of the *strip* and *miracle miles*;

– settlements formed synchronically, resulting from a critical and modern design. These are settlements like, for example, regional and superregional malls, but also settlements with other specializations (tertiary, productive, etc.). In a city like the one promoted and built by the modern movement, dominated by functional segregation, a specialized fabrics/settlements plan and design is an essential element. In the true reality of both European and American cities, the scheme becomes less abstract and the integrations with the traditional routes and settlement system more complex.

It is also possible to establish a division, very frequently seen in the scientific literature dedicated to commercial specialized fabrics (Coleman 2006, 64), but also applicable to other uses and scales, between:

– *In-town* specialized urban settlements/fabrics with urban localization, within the traditional/consolidated urban fabric (pedestrian mall, consolidated urban centers or downtown/historic centers' specialized portions for touristic and commercial functions). These settlements tend to be the result of a diachronic formative process, however, in the specialization phases a tendency for synchronicity can be noted;

– *Out-of-town* specialized suburban settlements/fabrics (e.g.: *regional and superregional shopping centers*), located outside the traditional consolidated urban fabric, in the suburbs; these settlements result much more frequently (except for the *strip* and the *miracle miles*) from a synchronic formative process[12]; they are located in antipolar positions in relation to traditional urban cores, and often act as a catalyst for the suburban expansion.

2.2 Fabric Scale

The aggregate scale analysis focuses on the complex inner organization within the settlements that polarize and constitute the nodes of the territorial network of routes.

---

[12]This grouping should also include specialized settlements located in rural areas or in areas with landscape value (e.g.: commercial districts located in rural non-urban areas of landscape value, combined with touristic centers, touristic villages and holiday resorts). These centers are also located in proximity of specialized route network nodalities, with a variable level of nodality from case to case. For the sake of limiting the research survey, in this book they are not studied and the attention is devoted only to commercial specialized settlement/fabrics, located in suburban areas.

In the previous section, the variety of settlements, possible combinations with variations in location and specialization level were taken into account. In this section, instead, the focus will be only on suburban context, specifically on *Out-of-town* commercial special settlements. The regional and superregional shopping mall is analyzed in this section at the aggregate scale; its component elements being identified and "re-named" in order to allow a typo-morphological reading and interpretation as a fabric.

Suburban commercial fabric represents a very high level of specialization. Other commercial fabrics, urban ones, such as the *strip,* downtown *mainstreet* or European shopping highstreets' fabric, unlike that of the mall, almost always include traces of the traditional "mixed" fabric, characterized by a prevalent residential function and progressive specializations.

Due to this high specialization level and to its synchronic formation, commercial suburban settlement fabric is a particularly interesting case study for reading the dynamics related to specialized fabrics. Such a reading of the suburban fabric with commercial specialization finds its application potential, as well as in the study of other specialized suburban, touristic, directional settlements such as those of the *edge cities* (Dunham-Jones and Williamson 2011, 172–200), even in the study of specialized urban settlements, commercial and/or more frequently mixed use, located within the historical and consolidated urban organism.

As for the territorial and urban ones, even at the aggregative scale, elements referable to the dyadic couple "routes and settlements/nodality" are identified,[13] along with several other component elements, roles and organizational hierarchies.

2.2.1 *Commercial Routes*

Aggregate scale reading in the specific case of the suburban mall constitutes a chance to identify elements, dynamics and hierarchies that, due to their similarity with those read in the fabric of the historic city, are particularly interesting for a crossed historical/contemporary fabrics' comparative reading perspective (Strappa and Buongiorno 2019).

Looking at the suburban mall aggregates, we can first identify a wide range of routes, which, each one in a different and proportionate way, all play the role

---

[13]Unlike what can be observed on the territorial scale—where it is possible to associate/coincide the settlements almost exclusively with the polarities/nodality—at the aggregate scale the settlement occupies and is built also on the fabric paths' pertinent strips, therefore the distinction between nodes/poles and settlement becomes particularly useful.

of a fabric generator[14]; these routes are polarized by further specialized building elements/units, which constitute the nodes/poles of the route network.

These aggregates are, in most cases, the result of a synchronic formative process. Only after some time, they become the object of more complex diachronic transformative process (Fig. 2.3).

As a result of the modern designer critical consciousness, these designed aggregates are generated through an accelerated process, almost like small experimental "in vitro" cities.

In any case, the strong influence that economic and commercial factors have had, the issues coming from behavioral psychology, marketing and social science disciplinary fields applied to commerce already in the formative process, constitute an important counterweight to the expressive freedom of the designer's synchronic critical consciousness. It is therefore possible to interpret in these fabrics a formative and transformative process that gives large spaces to "spontaneous"[15] components.

Several routes contribute to the special commercial fabric formation, with different hierarchical roles:

Main commercial route (matrix route): The open air or enclosed *mall*[16] is the route that generates the specialized commercial fabric, acting as its centralizing axis. Serial commercial units settle on its longitudinal edges. The *main commercial route* is polarized at its ends by *special nodal commercial units*, usually called *anchors* (Coleman 2006, 293–326; Strappa and Buongiorno 2019) in the specialized literature. It is also recognizable for being a loop circuit route. Other special commercial units, the *intermediate polarities*, are also located along the route, at the nodes. Intermediate polarities can be the result of merging several commercial serial units. A main commercial route has typical dimensions dictated by technical maximum distances between nodal commercial units, typically not exceeding 200 meters (Maitland 1985, 110).

---

[14]Reflecting on the relationship between road and commerce is fundamental for the interpretative reading of commercial settlements formative process. Since the first forms of commercial settlement on the road, up to the present-day suburban specialized settlements, even with changed means and materials, the settlement methods, which reflect shared human needs and uses, are quite constant and diachronic (Rudofsky 1969).

[15]The "*spontaneous*" component mentioned here, rather than referring to the disciplinary definition of *spontaneous consciousness*, is used to identify what is not attributable to the 'critical' component. For the notions of *spontaneous* and *critical consciousness* (Caniggia and Maffei 1979, 39–43).

[16]The word *mall*, before designating the main commercial path of urban/suburban fabrics with commercial specialization and the whole complex, designated the specialized ceremonial path, for parades, polarized by institutional buildings, typical of cities such as London, Paris or Washington. The character of a ritual/ceremonial path, which remains until the contemporary use, is strongly present from the beginning: the word *mall* derives, in fact, from *pall mall* game, an English version of the Italian renaissance game of *pallamaglio*, an ante litteram *golf* game that needed a play path similar to the monumental tree-lined and straight malls. (https://it.wikipedia.org/wiki/The_Mall_ (Londra)).

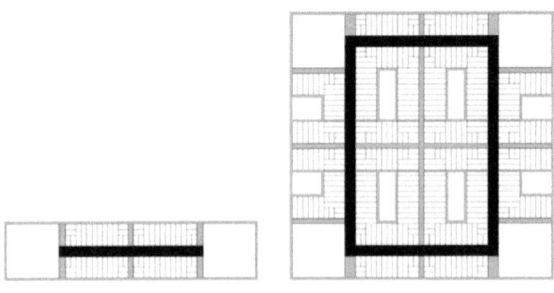

Main commercial route - matrix

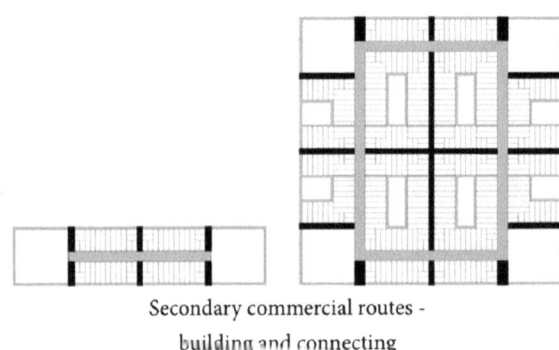

Secondary commercial routes -
building and connecting

Pertinent technical routes

Fig. 2.3 Aggregate's routes scheme *Source* V. Buongiorno

Secondary commercial routes (*building* and *connecting routes*): The distinction within the secondary routes is minimal; in the development of an aggregate characterized by a main route with a loop circuit, the secondary routes' role of *building* or *connecting routes* often tends to get confused (diagrams Fig. 2.6). However, it is considered useful to describe the characteristics of these routes. On the *secondary* *building* routes, the settlement and development of the aggregate takes place, further away than the settlement on a main commercial route.

These routes can be parallel or orthogonal to the main one. They are parallel to the main commercial route when there is an aggregate characterized by blocks that are not very deep in the direction orthogonal to the main route. Thus, on the routes orthogonal to the main one there cannot be the establishment of commercial units, so these orthogonal routes perform the function of *connecting route* between the main route and the *secondary building* commercial ones.

Otherwise, if the secondary *building routes* are orthogonally from the main commercial one, the aggregate is characterized by blocks deep enough to allow orthogonal building routes' that are repeated at regular intervals of about 100 m, also playing the role of fire escape routes.

The polarization of secondary routes consists, on the one hand, of the nodality generated at the intersection with the main route (and related *nodal special commercial units*) or of *polar special commercial units*. On the other hand, secondary building routes can also be polarized by the *parking polarity* (organized as one level or multistorey).

In the diachronic suburban commercial specialized fabric's transformative process, a secondary building route can be transformed into the continuation of the main route, where by extending the commercial fabric, the parking polarity is substituted with a special polar/nodal commercial unit.

The secondary connecting routes play the role of connection between secondary building routes (in the case of secondary routes orthogonal to the main route) or between secondary routes and main route or other polarities (very often dedicated to parking, as in the case of secondary routes parallel to the main route).

Pertinent technical routes: These routes, subordinate to main and secondary routes, are independent from commercial pedestrian traffic and constitute the dividing lines of the commercial aggregate.

They can be read as the overturning of the main and/or secondary routes within the commercial block. On pertinent technical routes there are technical areas pertaining to each serial special nodal/polar commercial unit; these technical "pertinent areas" are places for technical functions and for goods' storage. The pertinent technical routes' network is polarized by the goods' loading/unloading areas and by larger storage warehouses (Coleman 2006, 408–413).

Vertically/three-dimensionally turned routes: A suburban aggregate route network is developed horizontally and also three-dimensionally, offering a range of situations that goes from the malls typically developed on two or three floors, up to the Chicago *vertical malls* (Coleman 2006, 167–176) or the Hong Kong's mixed combinations of complex vertical fabrics.

The vertical or three-dimensional[17] development takes place through a processual transformative dynamic typical of urban fabric. The three-dimensional overturning of the main horizontal route can be observed both horizontally and vertically. The horizontal overturning, analogous to that which underlies the transformation and specialization of urban row houses fabrics to form the first special palatial and/or conventual-monastic complexes (Strappa 2014, 64), combined with the vertical overturning, is at the base of the *galleries* type, formed through "*passages*" specialization.

Vertical overturning has a more recent history, dating back to the introduction of the *Otis* escalators (Chung et al. 2001, 337–369). The vertical continuation of horizontal routes through the use of traditional stairs presented the limitation of having sections of the route (e.g., landings) that, although they were rendered homogeneous to the original horizontal route, through generous dimensions and continuous paving, they still represent interruptions, pauses between the main route and the overturned ones that therefore end up playing a subordinate role in the hierarchical order.

The overturning and vertical continuation of routes, when a lift is introduced, undergoes a great diffusion in the special building types for offices, but in commercial buildings instead it has limited success. Although the effort necessary to ascent the traditional staircases has disappeared, the discontinuity of ascent remains. In fact, even considering the possibility of a transparent or semi-transparent cabin that allows visual communication between inside and outside, or a casing that makes audiovisual interactivity possible from within the cabin environment to the outside,[18] in any case the horizontal routes located at the upper floors, resulting from overturning, remain subordinate to the route located at the entrance floor.

Even the location of special nodal commercial units (e.g.: *anchors* in department stores) can hardly be done on just the overturned floor/route: no matter how strong their polarization potential may be, they can only work if developed over several levels, on the entrance level and on those of the overturned routes. Thus, due to the need to be served not only by vertical ascent routes shared with the fabric, these special nodal units equip themselves with their own vertical routes, which run through the various floors. These vertical routes, due to the hierarchical importance within the fabric of the nodal unit that they serve, end up taking on the character of semi-public routes throughout the aggregate.

---

[17] As in other cases (special types such as the *palazzo*, the monastery, etc.), we face not only a vertical overturning, but instead a three-dimensional one: the original horizontal route of the ground/access floor extends vertically, (via elevators, stairs) or in a sloping line (through escalators, inclined *tapis roulants* or fixed ramps, inclined elevators), to then split and develop horizontal route networks to each upper level. The recent introduction, still untested in commercial fabric, of elevators with a complex three-dimensional path is the confirmation of the vocation to three-dimensionality in the overturned path dynamic.

[18] A good example of this interactivity is represented by the super-fast ThyssenKrupp elevator that stands in the One world trade center building in Manhattan, New York. The cabin's internal vertical surfaces are made up of video screens that re-create, by reproducing an audio/video story, a diachronic relationship between the city of today and its formative process of creation. https://www.youtube.com/watch?v=cKTPaqbXrAY—consulted on 20/09/2019.

Special nodal units have constituted the fruitful experimental field for the introduction of a vertical overturned route created with the use of escalators, particularly effective in the commercial sphere.

This mechanical route eliminates, like the lift, the fatigue of the ascent, but at the same time allows greater environmental continuity with the served spaces at each floor, and above all offers a perpetual mechanized movement that favors the *pedestrian flow* to the various floors, thus considerably reducing the subordination of the floors with overturned routes toward to the floors with main/entrance routes. We can consider it as the first real overturning/continuation of the main/entrance routes. The first experimental applications of this dynamic took place in department stores (Chung et al. 2001, 346–355), which almost always constitute the nodal special units within suburban malls.

To frame the vertical/three-dimensional routes in the global hierarchical order of aggregate routes, they can be defined as connecting routes when they are discrete/not continuous. This happens when the overturning is carried out by using elevators, but also, albeit to a lesser extent, escalators or *tapis roulant*. The latter two in fact, as well as the former, do not allow the establishment of commercial units on their pertinent strips.

Vertical/three-dimensional overturned routes act, instead, more and more as *building routes* when they are characterized by a certain continuity. This happens, for example, with the ramps (minimal inclination) that present commercial units established on their pertinent strips (e.g.: Mercado Libertad—S. Juan de Dios in Guadalajara, Mexico, Arch. Alejandro Zohn, Zohn 1999).

The overturned ramp route can play the role of the main commercial route if it consists of a ramped path with commercial units settled on its pertinent strips, polarized by nodal commercial units. Otherwise, a route that is vertically flipped with a lift cannot be considered as a main route (e.g.: Vertical mall Chicago).

Finally, it is possible to identify overturned vertical technical routes, similar to the horizontal ones, linking vertically the horizontal technical routes located on the various floors.

2.2.2 Commercial Nodes

The nodalities at the aggregative scale consist of:

Special nodal/polar commercial units, also referred to as *anchors* (I.C.S.C International Council of Shopping Centers "Shopping Center Definitions", https://www.icsc.org/news and-views/research/shopping-center-definitions) in the sectorial scientific literature. The denomination suggests their role of attraction and anchorage of the flow of pedestrian traffic, carried out in order to guarantee the commercial life of the whole aggregate. The number of commercial nodalities/polarities can vary, from a minimum of one (e.g.: Northland center, Detroit, USA - arch. V. Gruen), passing through the classic *dumb bell* scheme, with two large commercial surfaces that polarize the main commercial route, up to complex layouts characterized by more than two polarities on several floors (e.g.: Chicago Vertical mall, Hong Kong complex mall). These nodalities/polarities can be identified with special buildings, developed on several floors and equipped with their own vertically overturned routes (escalators, elevators, etc.). Due to their attractive power they act as public spaces (less public than the main route but more public than the serial commercial units) able to interject the pedestrian flow, channeling it in their vertical routes. In addition to performing the vertical connection function, in autonomy or complementarily with the vertical dedicated aggregate's routes, often the nodal/polar commercial units make up a complementary access route to the whole commercial fabric from outside[19] with their, albeit not always present, direct access from the external parking area.

Special non-commercial nodal units, these units have entertainment, leisure and public meeting functions (e.g.: the *food court*). The specific role of these nodal units is becoming increasingly relevant: they contribute to re-creating a simulacrum of urban life in an aggregate that does not otherwise favor it due to its high specialization. The importance of this kind of nodality/polarity in reality can be observed by looking at examples realized in the United States in the years between 1990s and the early 2000s. *Mall of America* or *West Edmonton Mall*[20] stand as pivots in the process of transformation of the mall type by putting not the goods' exchange activity at the center of the aggregate. In response also to the purchasing dynamics transformation phenomenon due to the advent of *information technologies* and digital *e-commerce*, the shopping center becomes mainly a place for experiences and urban life and then secondarily a place for goods' exchange (on contemporary and future mall's transformation, see Chap. 4).

Furthermore, nodalities consist of:

Non-commercial service units for the location of services (hygienic, technical, etc.);

Vertical routes cores, located at the intersection of horizontal and vertical routes;

Parking units and intermodal exchange nodes (subway/car/pedestrian mode) placed in the aggregate where the way of traveling along the routes changes.

---

[19] Access and free direct communication with the external parking lot, in the first malls and in the typical mature type, spreading from the 1960s up to the early 2000s, is not present. The entrance to the commercial aggregate is carried out through connecting routes, from the parking/railway station/intermodal exchange polarity up to reach the main or secondary commercial route.

[20] Both cases were developed by the Triple Five Group, a Canadian mall's development and design firm for whom non-commercial activities play an important quantitative and qualitative role. Other projects by the same firm: *American Dream Mall* in the metropolitan city of New York, and *American Dream Mall* in Miami, Florida.

2.2.3 Other Fabric's Elements: Block, Contrada[21], Base Serial and Aggregated Units

It is necessary to define also the (intermediate) component elements and their specialization for an understanding of the aggregate:

Commercial block

Aggregate formed by the commercial units (serial base units and/or special nodal/polar units), grouped in rows, on the commercial routes' pertinent strips.

The block includes the commercial units that contribute to form it, divided inside into sales area and pertinent area (warehouse + services of the commercial unit + technical compartment).

The suburban mall's commercial block is almost exclusively "basic" or weakly specialized: commercial units establish themselves on the commercial routes' pertinent strips, and expand to almost completely fill the "internal courtyard" space.

The presence of technical paths in the "internal courtyard" space or underground, for the unloading of goods in correspondence with each commercial unit's pertinent area, represents the first stage of specialization, which differentiates this block from a typical row base block where the pertinent areas are contiguous, but distinct and unrelated (Figs. 2.4 and 2.5).

Commercial *contrada*

The *contrada* is the first degree of specialization of the commercial fabric. In a fabric organized as a *contrada*, this specialization involves the routes, not the block. Each commercial row unit (basic serial or special/nodal) is recognizable as pertaining to the route on which it settles, regardless of the block to which it belongs. Since the suburban mall aggregate's blocks are almost exclusively basic blocks—or at least characterized by the absence of specialization dynamics that leads to the formation of a special nodal compartment in the inner "courtyard" space of the block—we can speak of commercial *contrada* in suburban mall's aggregate, as well as in historical urban fabrics (Caniggia and Maffei 1979, 136).

Serial commercial unit

It is the base unit of the aggregate, characterized by a small front on the short side and by a generous development in depth, similar to the row house that forms the traditional fabrics of European cities.

The front has a typical plan development of about 6/7 m (*frontage module*), the depth is variable but generally does not exceed four times the front's size (Coleman

---

[21] The italian term "contrada", refers to the system of specialization that characterizes european traditional urban fabrics, generated on routes' pertinet stripes. The contrada system is hierarchized by the generative route, ignoring the urban block (Caniggia and Maffei 1979, 136).

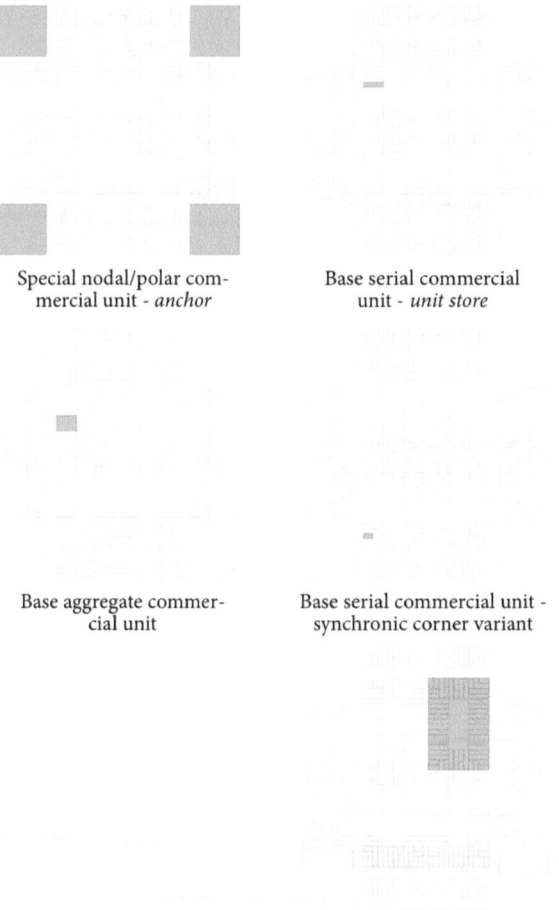

Special nodal/polar com-   Base serial commercial
mercial unit - *anchor*   unit - *unit store*

Base aggregate commer-   Base serial commercial unit -
cial unit   synchronic corner variant

Commercial block

Fig. 2.4 Commercial aggregate's elements *Source* V. Buongiorno

2006, 294). Exceptions are the synchronic variants, relating to specific conditions of the aggregate's development.

Aggregated commercial unit

This is an intermediate commercial unit, deriving from the merging of two or more serial base unit modules. The front of an aggregate unit ranges from 12 to 21 m while the depth can reach up to 30 m.

Fig. 2.5 Commercial aggregate's *contradas Source* V. Buongiorno

2.3 Building Scale

Within the commercial unit it is possible to recognize the dyadic elements (routes, nodalities/polarities, etc.) and the hierarchy that organizes them. It is, however, necessary to divide the description of these elements (and related internal micro-fabrics), referring to different building organisms. Specifically, the distinction is between base serial/aggregate commercial unit and special nodal/polar commercial unit.

2.3.1 Base Serial and Aggregated Commercial Unit

The surface of each unit is subdivided into: sales area, accessible to the public and to sales staff, and pertinent area, for goods' storage and technical functions, accessible exclusively to sales and authorized staff.[22]

The sales area is characterized by the presence of routes on whose pertinent strips display elements can be aligned. Special display elements are found instead at the nodes. This organization is very diffused in the actual retail system, dominated by the self-service system.[23] In this first classification phase of the book, we refer to a recent phase of the evolutionary process of commercial spaces, dating back to the twentieth century.

2.3.1.1 Routes

Main (loop) route

It is a special route that tends to perform as an obligatory route.[24] The purchase route is defined by the exhibition "scenes" formed by the goods' display units on its pertinent strips. It has a circuit layout, starting at the entrance of the commercial unit and ending near the same entrance, which also functions as an exit, in contiguity with the cash desk nodes. This route is also polarized by attractors positioned at the bottom of the business unit, consisting of special goods' display polarities (e.g.: meat or fresh merchandise department in supermarkets) or by additional cash desk nodes, or even areas dedicated to changing rooms. Other internal nodalities of the base commercial unit are represented by special nodal display elements ("island" display units horizontally developed), different from the base display elements (usually serial, vertical and back-to-back shelves). It is interesting to reflect on the analogy that this circuit scheme has with the routes of museums or amusement parks: in all three cases we face a specialized aggregate and a loop route. The loop route trend recalls to memory the shape of the bypass, although in these situations the need to avoid other traffic flows (even considering the technical flow of the goods transportation inside the unit) is missing. The loop of the commercial unit's internal micro-fabric is much more similar to an urban route interjected and continued inside the unit, generating and hierarchizing its internal micro-fabric (Fig. 2.6).

---

[22]Goods' storage spaces located in specific areas, in proximity to the dedicated technical routes, hosting services for the functioning of commercial units (bathrooms, changing rooms, etc.).

[23]The configuration prior to the diffusion of the *self-service* system includes a further subdivision of the sales area (which does not favor the interior "fabric" behavior) between the area occupied by the seller and the other one used by the customer. The division is represented by the desk/cash nodality—polarity, placed usually perpendicularly to the entrance (Chap. 3).

[24]Examples of loop route are found in complex commercial unit such as supermarkets and other specialized stores (Tiger store, Ikea store, etc.).

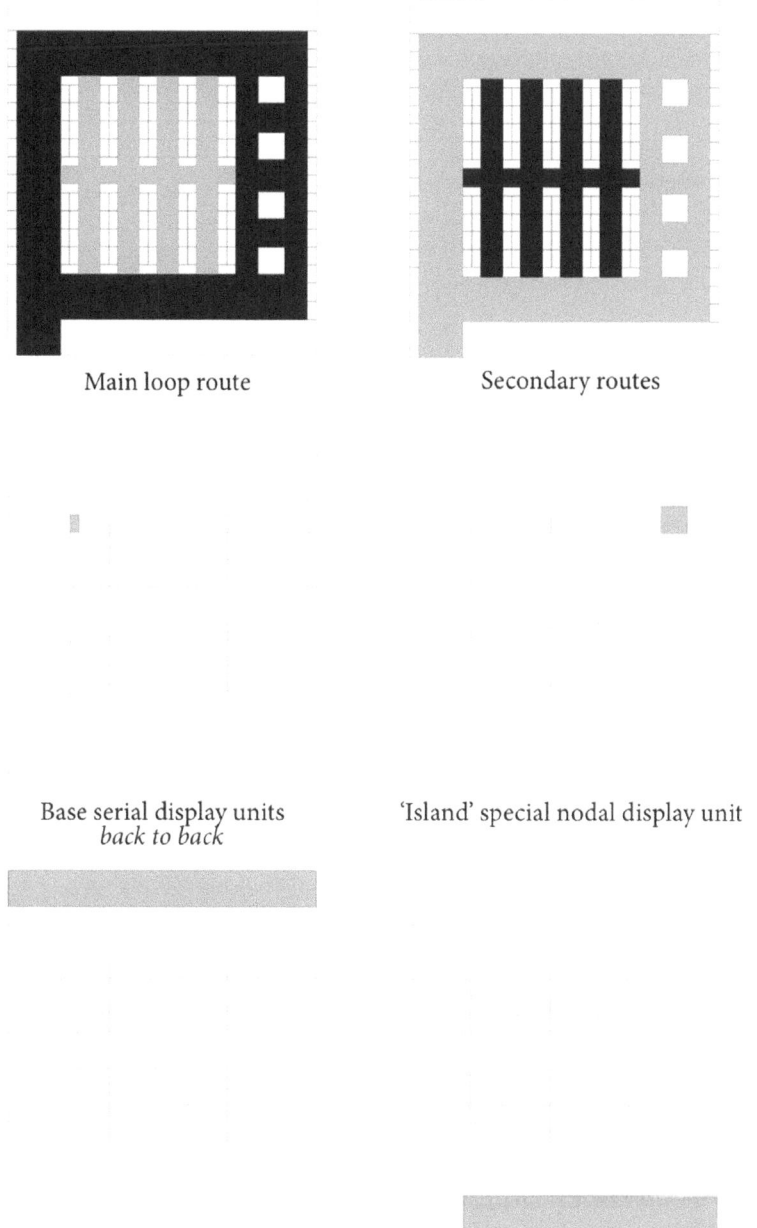

Main loop route   Secondary routes

Base serial display units   'Island' special nodal display unit
back to back

special goods nodality   Payment nodality. Access+exit

Fig. 2.6 Base serial and aggregated commercial units' routes and elements *Source* V. Buongiorno

Secondary routes

With the main function of internal distribution, these routes also perform the function of technical routes for workers in the operations of unloading and exhibiting goods.

2.3.1.2   Nodes

Within the internal micro-fabric of the base serial commercial unit, it is possible to identify the following nodalities:

- Cash desk node[25];
- Changing rooms areas, merchandise experience;
- Nodal or polar special goods' display areas (islands for displaying discounted goods or for window dressing);
- Nodal or polar areas devoted to giving information about goods (e.g.: cavea/lounge area in the Prada Epicenter store in NYC, architect R. Koolhaas—http://oma.eu/projects/prada-epicenter-new-york).

2.3.2   Special Nodal/Polar Commercial Unit

These are commercial units that polarize the fabric's routes, and in most cases, they coincide with the anchors. Their special character brings to them a relative autonomy in relation to the fabric to which they belong. They are characterized by the following aspects:

- from a structural and constructive point of view, special polar/nodal units typically have a different structural mesh, wider than that of the surrounding commercial fabric (Coleman 2006, 299–300);
- from a distributive point of view, they are characterized, in most cases, by a distribution hierarchized by a central point, which in some cases hosts a really special central nodal room with multiple heights (e.g.: Grand Magasins du Printemps, Paris); in other cases this kind of unit is polarized by the ascent/vertical routes nucleus (e.g.: New York's historic department stores such as Saks 5th Avenue or Bloomingdale's);
- the special nodal commercial unit, in its internal fabric organization, is characterized by the same elements of the base commercial unit; the main difference being the development on several levels and the presence of a special polar/nodal compartment and an internal node for routes' vertical overturning.

---

[25]The latest developments in the payment nodality transformative process are really interesting. With the advent of online payment (via app on a smartphone device), this nodality tends to disappear or to become an entrance hall to the commercial building internal micro-fabric. See, for example, AmazonGo, https://en.wikipedia.org/wiki/Amazon_Go—consulted on 09/16/2019 (Chap. 3).

2.3.2.1 Routes

The routes inside the special commercial unit are organized in a complex hierarchy which comprehends the following elements:

Main loop route

This is the main route of the special unit's internal fabric and it can be developed only horizontally (supermarkets, electronics stores or sporting goods) or through vertical overturning. It can be developed on several levels, with the formation of special rooms at multiple heights (e.g.: "Grands Magasins du Printemps", in Paris) or without (e.g.: department store in New York without a special room such as Saks 5th Avenue or Bloomingdale's). It is a route similar to that which characterizes the basic serial units and the aggregated serial units. It is polarized by the vertical/three-dimensional route's overturning nodes. It develops over the various floors through the elevator and escalator to reach the top floor, which due to its distance from the ground and relative discontinuity of path is usually also characterized by a polarizing, attractive and differentiated function (Figs. 2.7 and 2.8).

Secondary routes

These are distribution routes that intersect at the main loop route often orthogonally, near nodes and attractors located on the main route (cash desk nodes, goods' changing rooms, toilets) or in other portions of the internal micro-fabric (special goods display islands, goods changing room areas, nodal areas for information on goods and other internal nodalities).

Technical routes

Routes for loading/unloading and storage of goods partly distinct from the secondary routes and partly coincide with them.

2.3.2.2 Nodes

In the special nodal commercial unit's fabric, it is possible to identify the following nodalities:

– cash desk node;
– changing room areas/goods' sensorial experience node;
– goods' information areas (promotional/information stands or personal shopping service dedicated areas—as happens in large department stores, like the historic *Saks 5th Avenue*, in Manhattan, New York);
– toilets;
– special goods node: fresh goods, delicatessen, fish market, butcher, bakery, pastry shop, in supermarkets; discounted goods, limited series or sporting goods in department stores (Figs. 2.9 and 2.10).

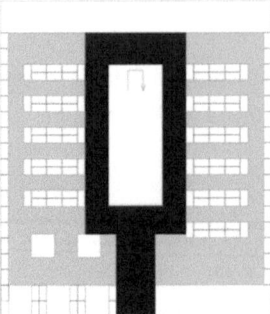

Main loop route

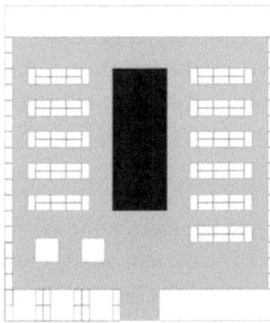

Multiple height special nodal room with overturned vertical routes (lifts and stairs)

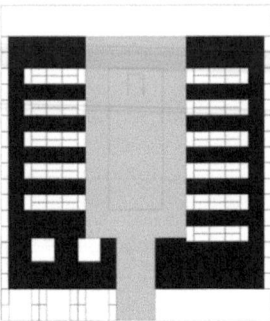

Secondary routes

Fig. 2.7 Special nodal commercial unit's routes *Source* V. Buongiorno

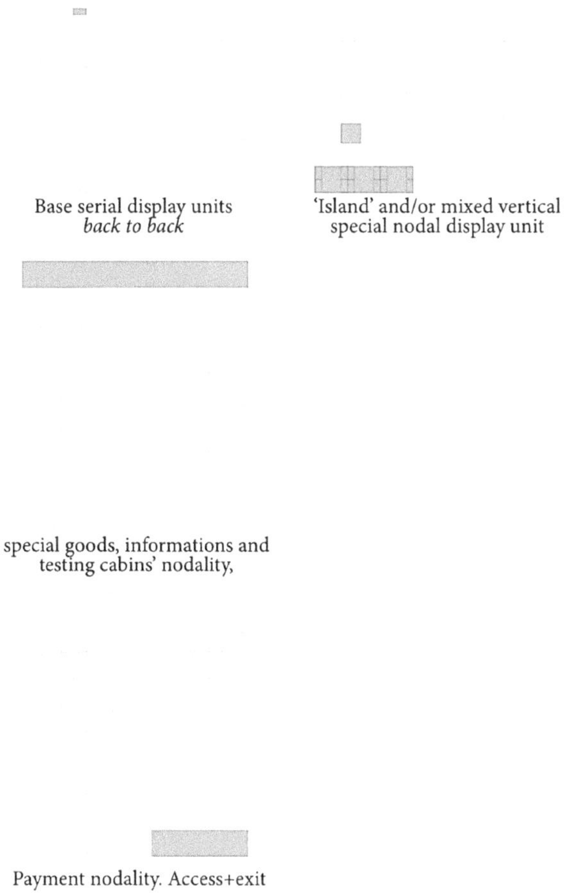

Base serial display units
back to back

'Island' and/or mixed vertical
special nodal display unit

special goods, informations and
testing cabins' nodality,

Payment nodality. Access+exit

Fig. 2.8 Special nodal commercial unit's elements *Source* V. Buongiorno

2.3.3 *Other Elements of Commercial Unit's Building Scale Inner Micro-fabric: Block,* **Contrada***, Base Serial and Aggregate Units*

In the interpretative reading of the commercial unit, in its complex nature of interjected micro-fabric, it is also possible to identify other fabric elements:

Internal micro-fabric's base block

It is formed by the base serial display units installed/established on the (main loop and secondary) routes' pertinent strips. The display units are arranged back-to-back;

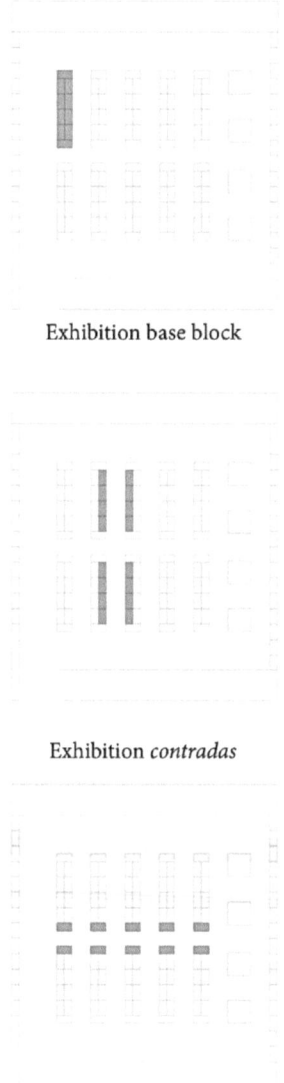

Exhibition base block

Exhibition *contradas*

Fig. 2.9 Base serial and aggregated commercial units' blocks and contradas

consequently, each block has the thickness resulting from the sum of the two serial exposure unit's depth. It does not present (following the analogy with the traditional urban fabric's block) adjacent pertinent areas located in depth behind the exhibition front (except for eventual display units characterized by double bottom and compart-ment for goods storage on the bottom), very often the pertinent area/temporary

Exhibition base block

Exhibition *contradas*

Fig. 2.10 Special nodal commercial unit's block and contradas *Source* V. Buongiorno

deposit is located on the lower floors, in the basement, at a height more appropriate for the depo than for the display.[26]

---

[26]Despite the multiplicity of variants, the organizational division, in the exhibition unit understood as a "storage furniture piece", between basement and elevation, confirms the hypothesis of a possible analogy between the traditional urban fabric and the unit's internal micro-fabric (Regazzoni Caniggia 1982).

Internal micro-fabric *contrada*

The special commercial unit's inner micro-fabric is organized, similar to the special urban fabrics' one, according to a system of *contradas* that follows a logic of merchandise/product specialization.[27] As in the urban fabric's contrada, each vertical exhibition unit is identifiable as belonging to a product lane/contrada (e.g.: home products lane, beverage lane, wines, sports clothing lane, clothing lane, perfumery, etc.) and as belonging to the route on which it is settled. In the basic block the fact is that a serial display unit that belongs to a block rather than to another has a secondary importance.

Base serial display units

Exhibitors developed in height, installed/established on the route's pertinent strips, positioned against the wall or leaning against each other, back-to-back.

Special nodal display units

Stand-alone display units, horizontal, for displaying special merchandise, discounted or characterized by other particular aspects;

References

Caniggia G, Maffei GL (1979) Composizione architettonica e tipologia edilizia: 1. Lettura dell'edilizia di base, Marsilio, Venezia
Chung CJ, Inaba J, Koolhaas R, Leong ST (2001) The Harvard design school guide to shopping/Harvard design school project on the city 2. Taschen, Koln
Coleman P (2006) Shopping environnements. Evolution, planning and design. Elsevier Architectural Press, Oxford
Dunham-Jones E, Williamson J (2011) Retrofitting suburbia. Urban design solutions for redesigning suburbs. Wiley, Hoboken, NJ
Hindley G (1972) A history of roads. The Citadel Press, Secaucus, NJ
Liebs CH (1995) Main street to Miracle mile. American roadside architecture. The John Hopkins University Press, Baltimore/London
Longstreth R (1997) City center to regional mall. Architecture, the automobile, and retailing in Los Angeles, 1920-1950. The MIT press, Cambridge, MA
Maitland B (1985) Shopping malls: planning and design. Construction Press, London
Regazzoni Caniggia A (1982) I mobili da riposto. Profilo di tipologia dell'arredo 2. Alinea, Firenze
Rudofsky B (1969) Streets for people. A primer for Americans. Doubleday & Company Inc., New York
Strappa G, Ieva M, Dimatteo MA (2003) La città come organismo. Lettura di Trani alle diverse scale. Mario Adda Editore, Bari
Strappa G (2014) L'architettura come processo. Franco Angeli editore, Milano

---

[27]In this regard, see the organization of the commercial fabric of the eighteenth-century *Baixa* district of the city of Lisbon, where each of the roads leading to the river shows in the toponym the memory of the merchandise/product specialization that has characterized them since the beginning: Rua dos Zapateiros—shoemakers, Rua Aurea—goldsmiths and jewelers, etc.

Strappa G, Buongiorno V (2019) Commercial urban fabrics updating. retail globalization and shopping cultural areas. In: Charalambous N, Zafer Cömert N, Hoşkara Ş (eds) CyNUM 2018 conference Urban Morphology in South-Eastern Mediterranean Cities: challenges and opportunities proceedings. Cyprus Network of Urban morphology, Nicosia,CY

Strappa G (1995) Unità dell'organismo architettonico. Note sulla formazione e trasformazione dei caratteri degli edifici. Edizioni Dedalo, Bari

Zohn A (1999) Alejandro Zohn Arquitectura Y Reflexiones. Unión Editorial, Guadalajara, MX

Chapter 3
Suburban Commercial Fabric Formative Process

3.1 Territorial and Urban Scale

Substratum—Traditional route and Le Corbusier's interpretation.

The development of specialized routes dedicated to a specific transport modality has only recently gained a significant acceleration. The railway, motorway and airway networks development has revolutionized a millenary tradition of transport and movement mainly based on human animal motive power transport (Hindley 1972, 2–6).

The traditional route, before its specialization during the industrial era, is characterized by a multiscale meaning and functionality. The most important routes simultaneously play multiple complementary roles, such as:

- connecting very distant areas of the territory, or even different geographical cultural areas;
- creating an urban structuring axis, by generating the urban organism and working as its axis;
- being a matrix route that generates urban fabric on its pertinence strips;
- being a matrix route at the building scale, as collector and generator of the inner micro-fabric belonging to buildings that are built on the route.

The generative role of routes in the historic city is clearly observable through the analysis of special building organisms' formative process, starting from the base fabric's transformation.

The traditional route, as well as the territorial, urban and aggregative organism of which it belongs, is also characterized by a progressive and reversible specialization process.[1]

---

[1] Examples of progressive and reversible specialization process: on the one hand, the process of recasting that leads to the formation of special building and urban organisms; on the other hand, the process of dequantification and subdivision typical of crisis periods, such as that of the Roman fabric's "medievalization" (Caniggia and Maffei 1979, plates 17–18).

© The Author(s), under exclusive license to Springer Nature Switzerland AG 2020   39
V. Buongiorno, *Suburban Retail Spaces*,
SpringerBriefs in Geography,
https://doi.org/10.1007/978-3-030-54991-6_3

In this organic universe, man's intervention, even at the smaller (building or interior space) scale, far from being a fragmentary action isolated in time and space, is the strongest act of critical transformation as interpretation of a process.[2]

The rapid route transformation and specialization process will lead to its isolation from the territory as will be analyzed in the following paragraphs. The gradual transition from a type of route acting as an exchange channel that makes fertile and feeds the territory on which it is grafted, to a more complex horizon made up of multiple layers, as many as there are specialized uses, then corresponding to multiple, overlapped and separated traffic flows will be clear.

Such a sudden and strong transformation inspired the lyrical and prophetic descriptions by Le Corbusier.

In the article *"La rue n 'existera plus"*[3] written for the French newspaper *"L'intransigeant"*,[4] he provides a precise and, at the same time, dream-like description of a new territory that renounces tradition.

Setting aside the traditional route, described as a "deep crack", a "narrow corridor" where the heart is oppressed, he guides the reader of 1929 into a new world.

A world where man is surrounded by trees, lawns, huge green spaces, healthy air and no noise, where tall residential and office buildings as crystals, distanced 400 m, set in the landscape in an unprecedented way. A metro station serves each one of this gigantic crystal buildings that act almost as indoor micro-cities, by connecting them in a grid pattern indifferent to car routes.

A world in which the routes show unprecedented three-dimensional developments that distinguish routes devoted to cars, trains and pedestrians. In this system the latter two, by using ramps, can walk through several terraces at different levels that hosts public commercial and civic spaces and related public life. By climbing this three-dimensional system, each pedestrian reach the top level where he can look at the new modern landscape made of pure prismatic volumes merged in the green of lawns and in the blue of the sky; overall, once gained altitude and aerial visibility, by looking down from afar on the millenary past of traditional urban fabric and routes, he can perceive the distant rumor of the ancient districts of Paris remained in their old crust.

The description by the French-Swiss architect, a very individual expression of critical consciousness, far from being only the fruit of his creative imagination, refers to a real territory that is already under construction. The first European experiments for creating specialized motorway routes, such as the Italian *Autostrada dei Laghi* (1922–23) (Moraglio 2007), and the contemporary *AVUS-Automobil-Verkehrs und Übungs Straße* in Germany, together with many highway projects contemporarily

---

[2] A global formative process in which, after the formation of routes that results from the recognition of the suitability of soil and orographic structure for the journey, the act of building as a further "finalized transformation of matter, can be read as a specialized modification of a part of the territory, specific technical-economic moment within the anthropization process: individual finalization of the operation, which will be followed by a collective finalization and an organic synthesis" (Strappa 2016, 25).

[3] "The road will no longer exist".

[4] *"La Rue: texte manuscrit pour l'Intransigeant 05/1929"*, Fondation Le Corbusier Archives. Corbusier and Jeanneret-Gris (1929).

carried out in the United States, as in the case of the Los Angeles region, form the reality palimpsest on which Le Corbusier's critical and creative interpretation is based.

These first realizations, transversal between two continents, introduce a new territory, where the street/*road*, as a complex abstract term and as a real place with multiple meanings, is progressively abdicated (for a deeper understanding of the concept of "street", see Rudofsky 1969; Anderson 1986; Rykwert 1995, 1982). After a process of abstraction of the route from the territorial organism, it becomes a mere traffic flow channel. This channel gradually specializes, due to needs of optimization of people and goods movement/transport, organizing itself within a complex hierarchical network, rich in internal differentiations, following the formative and transformative process that will be illustrated in the paragraphs of this chapter.

3.1.1 An Ancient Special Route: the Turnpike Road

The first and the most important characteristic that distinguishes the traditional route and its oldest specialized version is the split in the latter, of the route itself from the territorial, urban, aggregative and building organism.

With the specialization process, the dynamics of fabric formation on the route's pertinent strips, which had generated the majority of European and North American urban fabrics, gradually disappears, leaving isolation and impermeability among route, fabric and territory.

Back in seventeenth-century England, turnpikes roads were in use as toll roads. They were distinguished by the presence of buffer zones adjacent to the road, similar to actual specialized motorway routes' buffer zones, in which built structures and/or vegetation could not be settled. Traveler safety issues were among many other reasons at the base of such a restrictive regulation.

> …King Edward I had ordered the clearing of 200-foot swathes on either side of roads to remove the cover for lurking criminals; his grandson Edward III had repeated the injunction with particularly reference to the approaches to London (Hindley 1972, 72).

This measure is an attempt to avoid the frequent assaults on vehicles, travelers and goods made by criminals, the so-called highwaymen.

By limiting the chronological period of this research to the modern era,[5] it is therefore possible to trace back to the *turnpike roads*, as they are more directly connected to the contemporary motorway route type, a first level of route specialization, thus identifying a first phase of the specialization process.

In turnpike roads it is already possible to recognize some main features of the specialized route, still present in nowadays motorway route:

---

[5]Forms of toll roads have been built and used since ancient Roman times. The toll was very frequent near the bridges nodalities. During Middle Ages it was applied both to roads and bridges but above all to waterways, which experienced a new season of use due to the Roman road system decadence (Hindley 1972).

– discontinuity in relation to the surrounding territorial/urban/aggregative/building organism, both in plan (due to the presence of a technical buffer zone that isolates the route from the territory and transforms it in a dividing line) and in elevation/section (due to the presence of wide stretches of route made at levels different from ground floor, through embankments, trenches, tunnels or on bridges, to allow an easy slope useful for a faster and safer traffic);
– discontinuity in relation to traditional routes network. Access to turnpike roads is punctual and limited to nodal/polar positions (usually next to urban centers) marked by the presence of "toll gates" buildings, used for toll payment and entry/exit traffic control;
– last but not the least, a system of functioning, administration and maintenance of the specialized route, which tends to centralization in the hands of the turnpike trusts.

3.1.2   Continuous Specialization in Route Design and Construction

The development of the modern road surface, mutual influences between means of transport and route features.

Developments in road construction techniques during the eighteenth and nineteenth centuries brought changes, especially as regards the road surface constructive technique. The *Trésaguet* method, which later evolved into the *Macadam*[6] method, is both based on the superposition of layers of lithic material that is increasingly minute toward the top surface in contact with vehicles and atmospheric phenomena. They respond to the need for a more uniform and smooth traveling surface, able to allow a good draining, for a safer and faster road use.

The road surface constructive technique's refinement process will reach a mature stage, simultaneously with the first great wave of automobile use, at the beginning of the twentieth century. At this time, the compelling need for a compact road surface leads to the design and realization of many roads with bituminous surfaces, similar to those paved with *Macadam* system, but with the improving addition of compacting and waterproofing bituminous material that reduces the dust production and abrasion of mantle material during fast motor vehicles runs.

Road surface specialization marks a pivotal moment to be interpreted within the broader motorway network specialization process horizon.

In the United States, at the beginning of the twentieth century, the creation of road samples by using avant-garde bituminous paving is made to offer and spread the experience of new possibilities for motorists; at the same time, it wants to promote and fundraise for the construction of new highways, such as the *Lincoln Highway* (Liebs 1995, 17–19).

---

[6]The methods' names derive from the homonymous engineers who developed them: Pierre-Marie-Jérôme Trésaguet (1716–1796) and John Loudon McAdam (1756–1836).

This specialization process leads to an optimization of route's structure also from the planimetric and sectional point of view (Fariello 1963). Transformation in planning responds to the need for specific curves and curvature radius, congruent with the technical characteristics of the motor vehicles and their achievable speeds. The same question applies to the section design, where the optimization attempts, as far as possible, to eliminate large level differences and slopes that are too inclined, through the construction of route sections in embankment, in trenches, on viaducts and/or in underground tunnels that allow constant slopes, ideal for modern motor vehicles' circulation.

So, the new route "bypasses" the territory and is indifferent to its orographic conformation, as well as being shaped as a specialized traffic flow channel that is abstract and separate (at least in its continuous portions) from the territorial organism to which it belongs. It appears extremely different if compared with traditional routes, which instead lived in a close solidarity with the orography. A solidarity that represented the heritage of ancestral anthropic acts such as that of orography's interpretation and recognition of the orographic palimpsest susceptibility to be transformed in some of its specific points and lines (e.g.: ridges) in paths and later routes (Strappa 2016, 26).

The only surviving relationship between specialized highway route and territory is merely visual, vitiated by the in-speed perception and by critical landscape design interventions on the contiguous areas, developed taking into account the motorist's "fast" and detached fruition (Hindley 1972, 61).

3.1.3 Nodes and Poles Formation for the Special Routes Network

The spread of automobile use and its becoming a private consumer good, together with an exponential increase of the number of vehicles in circulation—firstly in the United States from the 1920s and then in Europe during the second postwar period—marks the beginning of a further specialization phase of special routes. The phenomenon of quantitative growth of the number of vehicles is a consequence and, at the same time, also a cause of qualitative changes of built environment's structure. Suburban city model diffusion, which indeed had already seen a first development with the rail-tram network development during the nineteenth century,[7] experienced

---

[7]The spread of the rail and tramway network has strongly structured the bourgeois expansion of the nineteenth-century city, both in Europe and in North America, and given impetus to the development of the first suburban settlements arisen around the nodes of the collective transport network, fundamental for the movement of the masses of workers and the industrial bourgeoisie. The first American and European *banlieue* is a variation/imitation of rural/suburban dwellings of the upper classes, with the important variant of the quantity of inhabitants and improved living conditions. Places for a bourgeois class in exponential growth, escaping from the unhealthy living conditions of historical centers and downtowns increasingly clogged by traffic and incompatible industrial activities, the new suburban settlements of the nineteenth century consist mainly of base

a season of great progress, through a mutual mechanism of production and influence, between suburban city and motorway infrastructure.

Route surface transformation and specialization, in plans, sections and in its constructive characters, allowing a safer use experience for fast motor vehicles, introduce an unprecedented perception of distances within urban and territorial organism. Those same distances that separate suburban residents from old and new downtowns of suburban cities like Los Angeles,[8] whose daily journey would become impossible without the massive car diffusion and use, are now justified and made possible by the contemporary and fast construction of an urban and territorial motorway routes network. As already mentioned, this network, as well as its component elements, is characterized by being abstract in relation to the territory, to its orography, and has contacts with the latter just in nodality and polarity points. These points are positioned next to existing urban settlements, according to a spontaneous logic, or in completely new nodes/poles, that according to a critical and modern design logic are planned to host new urban/special settlements.[9]

The specialized nature of the new routes' network implies that a high level of specialization characterizes and regards each one of its components, the routes as well as the nodal/polar settlements.

It is possible to distinguish two groups within nodal/polar settlements:

A. Nodal/polar settlements that derive from a diachronic specialization process, characterized by mixed functions and results from the evolution of pre-existing urban centers (e.g.: European historical centers, North American downtowns). Some examples belonging to this group of nodalities/polarities, as regards commercial specialization, are: pedestrian malls and pedestrian urban areas.

B. Polar/nodal (completely new) settlements that derive from a synchronic process, with a stronger exclusive commercial specialization; usually located in the new suburban expansion areas. They are built almost contemporarily with the related specialized routes' network. Some examples belonging to this group are: *strips, miracle miles, drive-in markets, regional and superregional shopping centers, edge cities,* etc.

In the following pages the process that leads to the formation of nodal/polar settlements will be synthesized for both groups.

A. The formation of specialized nodal/polar settlements, deriving from existing urban nuclei constitutes a sub-phase of a bigger and complex formative process.

---

building aggregates with some first necessity/daily facilities. The core of services and facilities, in this first phase of suburbanization, remains still in the historic city center/downtown. As *ante litteram* laboratory of the modern suburban suburbs, this first *banlieue* sees the birth of the *bungalow* building typology, which would have become the basic element of the suburban city of the twentieth century (Lacroix 2003, 75–80; Case Sheer and Stanilov 2004).

[8] Los Angeles during the first decades of the twentieth century represents a real experimental laboratory for the construction of the suburban city (Longstreth 1997).

[9] As regards urban nuclei refer to the urban downtowns/historical centers, which constitute the main polarities in the development of the network, or to the "new downtowns", arisen during the suburban expansion within the first peripheral urban ring and characterized by mainly commercial functions, such as for example the *miracle miles* or the *strip* (Longstreth 1997).

The first experiments on urban fabric specialization took place in the existing historic centers and downtowns. The need for a rational functional traffic flow organization is evident in the historic urban center streets, rather inadequate to host a high volume of vehicular traffic, such as that produced by the contemporary car-oriented society (Liebs 1995, 24–27). The most problematic situation is that of commercial urban routes such as the American *main street*, or the Italian "*Corso*". They are in the typical situation of territorial routes, which played an important role in generating the urban fabric to which they belong. Because of their higher hierarchical value, compared to other routes, they host specialized functions such as those for commercial activity and community exchange within the urban settlement. In pre-industrial times, the trafficking of horse-drawn carriages, men and other animal-drawn vehicles, while constituting an element of disturbance, and in a certain sense dividing, remains however a vital factor for the territorial/urban route and can well coexist with the commercial special function, because the quantity and quality of traffic is still acceptable and compatible with the mixed functions of the traditional urban fabric.

The pre-automotive territorial route, which becomes a *main street* in some urban sections, is pivotal for the territorial and urban organism life. While hosting a considerable volume of traffic, directed from a territorial polarity to another, it succeeds also in generating urban and building life on its own pertinence strips, producing building units and, through aggregation, urban fabric.

This situation of equilibrium in pre-industrial urban and territorial organism is questioned when a large number of motor vehicles start to circulate and above all, to stop/park, for purchase-related activities (e.g.: goods' loading and unloading), in the city center, clogging it up.

The further specialization of the special territorial routes' network, in its nodes, constitutes a decisive moment in the general process, specifically characterized by the implementation of bypass routes.[10]

For the smaller nodalities (simple intersections of two or more routes), already in the first European experimental realizations like the Italian Autostrada dei Laghi, the *flyover* is used, a bypass node on a smaller scale, which allows the independent and simultaneous transit of vehicles on different routes and directions, through overpasses and underpasses on several levels.

For larger nodes, located within existing urban centers, the more complex situation due to the inadequacy of traditional preexistent routes to host a new quantity and quality of specialized traffic, a new suitable solution was needed. It came from the experimentation with the transformation of existing urban areas.

Both in the American and European contexts, the construction of the specialized routes network went through the transformation and reuse of existing ones (traditional routes characterized by a multiple specialization and interscalar value and role). It was only later that the construction was made through the identification of new route tracks that could better fit to the degree of efficiency and autonomy/abstraction that the new specialized routes' system requires.

---

[10]It is significant that in the US road nomenclature, when it first appeared, the bypass is identified as a Special Route. (https://en.wikipedia.org/wiki/Special_route—consulted 25/08/2019).

The experiments carried out in the United States gradually and progressively develop an urban and territorial scale version of the specialized *flyover* road nodality. The vehicular *bypass* was the most used solution in the early 1900s to free the clogged city centers from the excessive traffic volume and parking spaces created spontaneously. But the deviation of traffic in a loop ring route centered on the freed route solved the problem only temporarily, by moving the traffic problem in a more antipolar area that slowly will become clogged up too. Others bypasses will follow the first one, to free the clogged former urban fabric bypasses, progressively, until reaching not urbanized areas (outside the urban perimeter) where it is possible to trace and build a totally new and independent non-urban special bypass driveway. Thus, gradually the urban centers and the portions of the old territorial routes comprehended within them are bypassed, through the creation of progressively larger rings (the firsts obtained using routes of the urban fabric, the others by tracking new special routes), until defining new autonomous tracks. Thus building a sort of horizontal stratigraphy like that reported by C.H. Liebs:

> ...the history of the American bypass is well illustrated in communities throughout the country. A parallel road to Main Street is laced with the remains of elderly filling stations; another road farther out of town is lined by once-prosperous motels whose peeling signs now advertise "Rooms by the week"; and farther out still is a much wider and busier thoroughfare where McDonald's and Holiday Inn have set up shop — all are physical reminders that the game of bypass leapfrog had once been played in town (Liebs 1995, 26).

In North American context the development of the network and its specialized nodes takes off rather rapidly during the first half of the twentieth century, and in the European, specifically the Italian one, the creation of a specialized motorway routes' network and the specialization of the existing nodalities take place simultaneously and lasts for a much longer time.[11] After the successful experiment of the *Autostrada dei Laghi* in 1920s, the world's first road equipped with multilevel nodes, tunnels and flyovers (Hindley 1972; Moraglio 2007), conceived as a special route aimed particularly at touristic use, a real and wider special network development started only after the Second World War.

In the 1960s, during the economic boom experienced by the Italian economy, a plan for the construction of a highways network started (network that includes the first built, iconic *Autostrada del Sole*), in the perspective of connecting the still very young national unitary state. Finally, the specialized bypass nodalities development process has undergone a really fruitful phase later, in the following decade. Since the 1970s several urban/territorial scale bypasses, such as the Roman *GRA—Grande Raccordo Anulare* bypass ring road, have been built.

B. But initially, specialized bypass nodes are only partially used for the realization of external not urban nodalities/polarities. The incoherent urban fabric development on dividing specialized routes' pertinence strips (following rules of traditional fabric

---

[11] A common aspect of both the European and North American contexts is the development of the motorway network, at its beginning, by private initiatives. The construction of demonstrative sections by private companies has started a development process in which the public administration, later, becomes the main actor (Liebs 1995, 18; Hindley 1972, 98).

in a new and different environment) will be practiced for a long time, especially in the areas of new suburban expansion. The commercial *strip* or the *miracle miles*, despite being built on highly specialized routes, such as *highways* or *urban motorways*, continues to generate a fabric on the strips belonging to these routes, that is commercially specialized and has strong problems of clogging, lack of parking and above all segregation of the various sections of urban fabric that gravitate on the route itself due to its being a dividing line and an impermeable barrier.[12]

This anomalous dynamic of nodalities/polarities development must be read in the context of the wider suburban city formative process.

The availability of individual movement by car, together with the low prices of land in areas that previously were too far from the urban center to be urbanized, and suddenly became precious for the car-oriented real estate market, favors the birth of new settlements, predominantly residential. They are characterized not only in North America but also in Europe, by the use of *bungalow* building typology, an isolated single-family house located in the middle of the plot.

The introduction of the new suburban settlement model marks a pivotal passage in the urbanization process that until then, even in the construction of the first suburban settlements on the railway lines and/or on the first urban highways, had followed a continuous and constant process of growth starting from a route that becomes a fabric generator.

The new special route becomes a problematic generator for a contradictory fabric and, due to its highly specialized nature (highway or railway), stands as a *dividing line*.

The *strip* is very representative of this process, being a fabric that rises, initially only residential then commercially specialized, on and from the new highways specialized routes' pertinence strips.

A similar situation is that of the *miracle miles*, so called by referring the original *Miracle Mile* complex that arose in the first suburban expansion of Los Angeles. These settlements, which later will become an important experimental laboratory for the mature special commercial fabric formation, inherit the logics and contradictions of the pre-existing urban fabric together with the clogging problems (Longstreth 1997, 127–141).

In an attempt to find a solution to this structural contradiction, and to the consequent problems of congestion (similar or even worse to those of historical downtowns and pre-existing suburban expansion), suburban development will lead in the second half of the twentieth century to the widespread diffusion of "*cul de sac*", a settlement model that is less anchored to the route and is capable of distributing uniformly the residential buildings/bungalows by isolating them in the center of each plot, freeing up the routes' pertinence strips. Inspired by the English *Garden City* transplanted in

---

[12]It will be possible to see, later on in the book, how the further specialization of the *strip*/*miracle mile* route, through the creation of a bypass and its pedestrianization, will have a fundamental role in the suburban mall's formative process. At the same time, in pre-existing urban centers/downtowns, the specialization of the territorial route in the nodes is pivotal in pedestrian malls' formative process, especially after the 1970s oil crisis.

America, it was first experimented in the New York region, at Levittown, NY, in the early decades of the century (Vachon et al. 2004).

Specialized nodalities/polarities formative process passes, indeed, through an internal process of specialization and transformation of what is built in modern times but with ancient logics.

For example, the *Miracle mile*, in a gradual transformative process (in the critical consciousness of numerous architects and in their designer and built projects), through its specialization—with the central route, ancestor of the current *mall*, that progressively specialized for pedestrian—gives origin to the mature suburban mall's commercial fabric, conceived as a nodality/polarity of the routes' network but at the same time is a considerably autonomous and specialized component element. This specialization process includes the creation of special bypass routes, similar to those used for downtowns and historic centers. As a result of this process, we arrive at a new type of nodality/polarity, such as that of *regional* and *superregional shopping centers* (that refer only to commercial specialization field). Widening the horizons it leads to *Edge Cities'* fabric (Dunham-Jones and Williamson 2011, 172–20), characterized by mixed specializations, or to touristic/naturalistic polarities/nodalities, all characterized by a high level of specialization, autonomy and isolation in relation to the territorial and urban organism (For a more complete discussion on the specialized fabrics that make up the nodalities of specialized route network, see Sect. 3.2. Aggregate scale).

3.2 Fabric Scale

Substratum—Traditional urban commercial/mixed use fabric.

To focus on formative dynamics at the aggregate scale, it is necessary to introduce the fundamental characteristics of the traditional urban fabric, here considered as substratum of this formative process.

Traditional fabric, as for example that of the modern Rome, is characterized, like its generator route, by a very high flexibility and metamorphic capacity. Its component elements are capable of progressive specialization, through transformative dynamics, recasting and knotting. They are also capable of de-specialization, through dynamics of *de-quantification.*[13]

The commercial specialization within traditional urban fabrics, both European historic centers and American downtowns, does not exclude the coexistence of base housing and special commercial functions, as in the raw house building type (specifically in the "shop house" synchronic variant) (Caniggia and Maffei 1979, plate 16A; Strappa 1995, 230). This integration of functions characterizes so much special construction. It is useful to consider, for example, what happens to the ancient Roman *domus*, during and after the 'tabernisation' phenomenon, or, in modern times, the

---

[13]Process similar to that of "medievalization" (Caniggia and Maffei 1979, plates 17–18).

presence of commercial spaces on the ground floor of some Roman renaissance buildings such as Palazzo Caprini, also known as "Raphael's house" (Chung et al. 2001, 30). Several economic, social and political factors, and above all, the limited possibilities of territorial movement in pre-industrial times, make it necessary to allow inhabitants the access to base functions (residential) and special ones (commercial, religious, entertainment, etc.) in the same area at a walkable distance. In this situation the route, the element at the origin of fabric's formation, acts as a *centralizing axis* (Strappa 1995, 82), the very core of fabric's life.

In the early phases of the fabric's specialization process, before the formation of nodal building organisms, with their own special nodal room (such as the *Palazzo*, the Church, the Convent, etc.), it is this centralizing route that takes on the value of a special space, being the subject of a first form of fabric specialization, that of the *contrada*.[14]

In the *contrada* each one of the "building individuals" settled on the urban fabric's routes on their pertinent strips is recognizable only for belonging to the route on which it overlooks. It is much less important to belong to each urban block. Thus the "special space" of the route, the central axis, is the real identity place of the fabric. It can be characterized in its development by one or more kinds of specializations along its pertinence strips (e.g.: commercial product specializations, residential social groups specialization, etc.), as readable in the toponymy of some spontaneous historical fabrics or in some modern, critically designed fabrics (e.g.: the eighteenth century "Baixa pombalina" district of Lisbon, whose routes' toponyms, until recent times, preserve the commercial product specialization of the initial project).

The introduction of new transport means, such as the railway, but above all the massive spread of car use, distorts this condition of relative balance in the traditional fabric. The possibility of rapid mechanized movement allows the concentration and segregation of special functions in polar positions, in a way that is clearly higher than in the past. The constraint of pedestrian usability disappears, the latter becoming more and more an obstacle to the majority of fast mechanized movements.

This functional concentration and segregation is accepted as a response to traditional historical fabrics' traffic congestion problems, consequent to the increase in the number of personal vehicles in use during the first half of twentieth century within the old and unsuitable network of traditional routes.

---

[14]"…*intendendo per contrada sia il percorso, che le case che vi si affiancavano nelle due simmetriche 'fasce di pertinenza'*".

notion of a *contrada*: "… meaning both the path, and the houses that were settled side by side in the two symmetrical" pertinent strips " (Caniggia and Maffei 1979, 136).

3.2.1 Formation of "Traditional" Urban Fabric on New Suburban Routes: Miracle Mile and Commercial Strip

In the context described so far, functional segregation is considered as a tool to govern an unprecedented urban organism with huge, unusual dimensions.

The car has become a mobile extension of living space, allowing individuals to live several miles away from workplace or leisure polarities. It allows the segregation of the residential function, more than the railway and in a more capillary way. Thus, for example in Los Angeles, the considerable flow of new residents, coming from internal and international migration, during the first decades of twentieth century (Longstreth 1997), will settle in new portions of the Californian city that are in development in those same years, large suburban subdivisions of the extensive American west coast territory, organized with very low housing densities and dominated by the new single-family bungalow housing building type.

The new car-oriented urbanization method is located at the antipodes with respect to the traditional fabric: the individuality of the bungalow (standing in the middle of each plot, detached from the perimeter) replaces the solidarity between building units that made up of the traditional aggregate-fabric; the monofunctional trend that leads to assign a single use or function to each plot or portion of land replaces the traditional complex integration between basic and special functions; the road ceases to be the special space of a *contrada:* in the new introverted system that considers routes only as technical channels for movement of goods or individuals among residence and special buildings, it can no longer play this role.

In cases of greater organizational complexity, such as in the "*cul de sac*" settlement scheme, an order with a higher level of hierarchy and organic coherence is readable between lots, housing units built areas, and the common lawn pertinent area. The large common lawn in this more hierarchical organization is readable as a virtual special compartment/room.

As already stated in the preceding paragraphs, suburban development on the one hand excludes non-residential functions from the fabric, and therefore also the housing-related commercial ones. On the other, it produces, during the first decades of its spread, contradictory cases of fabric generated by using the formative rules of the traditional fabric (such as the settlement of building units on the route's pertinence strips, the solidarity and contiguity between aggregated building units, the co-presence of basic and specialized functions) applied, instead, on the routes of the new suburban settlement system, which are incompatible and unsuitable, since they are dividing lines, for the formation and organic behavior of such a fabric historically based on accentrating axes.

The American experience of commercial *Strip* (Liebs 1995, 3–39) and *Miracle miles* (Longstreth 1997, 127–143), or even the contradictory development of some European and Italian suburbs that reproduce the traditional fabric on the unprecedented special driveways, although with different dimensions, leads to produce new traffic clogging pockets, repeating the situation from which the new suburban inhabitants fled, moving away from downtowns.

Both in the simplest developments, such as the *strip* dominated by basic commercial units, and in more specialized ones, such as the *miracle mile* characterized by the presence of the special commercial units like the department store, or even as certain commercial districts that give origin to the *drive-in market* and later the *supermarket* (Longstreth 2000a), the establishment of the aggregate still refers to an already anachronistic idea of route as a centralizing axis.

Due to the quantity and quality of the traffic flow that weighs on the vehicular special routes on which the new commercial aggregates settle, these same routes are incompatible with the organic functioning of an aggregate centered on them: from being the origin of the aggregation, they end up becoming physical and perceptive barriers to the exchange between the fabric's portions that gravitate on them. At the same time, their divisive and disruptive nature inhibits the formation of a more complex hierarchical structure and by consequence it also excludes the presence of a minimum level of fabric specialization, like the *contrada*.

The development of such commercial fabrics takes place in proximity to relevant nodalities of the system of specialized motorway routes, especially on territorial routes that connect downtowns with other important territorial polarities, and are located in the crown that surrounds the central core of traditional fabric.

Their formation receives an important boost when some of the most important and consolidated special business units, moving away from downtown locations, start to build new stores and offices in the suburban area to follow their customers moved to the suburbs. The location choice falls on the intersections of main routes, where it is possible to attract a much stronger flow of motorized buyers.

There is, therefore, the formation of concentrations, in few nodal points, of special commercial units, responding to a spontaneous market and competition law that wants them concentrated in the same place (Figs. 3.4 and 3.5).

Later in this special fabrics' formative process, as in the case of the Los Angeles *Miracle mile*, there is the almost simultaneous establishment of other special units and, above all, of basic commercial units, with smaller dimensions, in the free lots located between the special commercial units. In this way the fabric thickens its plots and increases its level of internal hierarchical complexity, also through the differentiation between special units (mostly department stores) and basic commercial units. At this point of the process the existence of a system of relationships within the fabric is very reminiscent of the traditional fabrics, characterized by the strong relationship between basic building units and special ones, and also anticipates the hierarchical structure of modern shopping mall type.

New commercial special fabrics responded to the previously unseen need for car parking areas for shoppers, by locating parking lots in-line, orthogonally or inclined along the edges of the driveway (curbside parking), or "off street" parking lots within the plots and facing the route (e.g.: parking lots in the drive-in market's "court", or in the space located in front of the commercial units' of some *strip* complex, or located in contiguous areas, as in some of the first *supermarkets* (Longstreth 2000a)). In any case, not only the presence of vehicular traffic but above all the centrality and importance given to car parking in such an aggregate, makes its organic pedestrian functioning difficult. The complete use of these aggregates by a pedestrian is almost impossible, first of all due to the difficulty to reach the special aggregate by foot, in a completely car-oriented urban organism, and then, due to the uncomfortable obligation to cross a double, if not triple, set of motor traffic barriers: the barrier of two parking types (areal and linear curbside) and the barrier of the traffic of vehicles transiting on the territorial route.

The new road section, enlarged to allocate the current traffic flow, together with the parking spaces, also produces a considerable distancing of the two facing commercial fronts, which face each other but do not exchange information or significant flows of buyers. Finally, even though the street and the aggregate are equipped with parking lots, because of the continuous progressive increase in the number of vehicles, the problem of clogging becomes worse. The same situation as downtown, from which inhabitants and merchants fled, is re-created: the simultaneous impossibility of using the territorial route both for those users who are not directed to the aggregate for purchases, and for those motorists who are instead going to commercial settlements and therefore find themselves with few possibilities of parking and pedestrian movement between the various commercial units.

3.2.2 Route and Aggregate Specialization: the Bypass Formation

The phase of development of the first specialized commercial suburban fabrics, with the emergence of traffic clogging-related problems, outlined in the previous paragraph, is followed by a phase in which the clogged situation of specialized suburban tissues is resolved with the same tools used for downtowns, through a progressive specialization process which leads, even in the suburban context, to the formation of *bypass* routes.

But the process of specialization of the new suburban fabrics is characterized by an intermediate phase in which, starting the specialization process, the parking areas are gradually removed from the main route, transferring them to the commercial units' backyard pertinent area behind them. This processual dynamic can be observed in the cases of *Broadway-Crenshaw center* and *Westchester business center*, both located in Los Angeles. In both cases there is a common problematic phenomenon related to the unbalanced use of the commercial fabric, with main access on the main route, still accessible to vehicles, and deserted rear/secondary access, spontaneously used as main access doors. This leads to conflicts between customers and goods delivery traffic flows. In this phase a first step of the transformation begins, through the progressive specialization of the main route that will become, through gradual pedestrianization, a *centralizing axis,* from being a *dividing line.* Simultaneously to proceeding in the specialization of the main route (that will gradually be transformed in the *mall*), the transformation process leads to the realization, albeit partial, of the bypass route.

The *Westchester business center* (1942), for example, demonstrates the partial adoption of the bypass route principle. On the route that delimits and serves the commercial portion of the fabric, there are large areas for parking vehicles.

The presence of the parking area that surrounds the commercial fabric and distances it from the new bypass route allows intermodal exchange and is a fundamental element of the specialization process. Thanks to these new larger parking areas located next to the bypass, motorized users can change their transport mode, from car to pedestrian, and then, after having completed their purchases, back to car mode. This intermodal change dynamic is similar/comparable to that one which takes place in a more urban context (mainly European) in urban fabrics characterized by a commercial specialization located next to the railway/underground transport network nodes.

In these first proposals that incorporate the bypass as an integral part of territorial and specialized aggregative organism, the centralizing specialized route nevertheless continues to offer the possibility of vehicular traffic. In contrast, the possibility of curbside linear parking is limited, if not completely excluded. The result is a centralizing route that has a representative role in the fabric, like for example in Los Angeles' *Westchester business center*. In this exemplary case, the customers access the specialized fabric with their vehicles through the centralizing route, faced by the commercial units' main façades of the commercial fabric. However, to use the intermodal parking/exchange, the customer must go to the commercial units' rear and from there, through secondary access doors, which become the spontaneous main access; finally they enter into the commercial unit. In this situation the main fronts/façade, with their display windows and official access doors, almost exclusively perform a scenographic welcoming function. Moreover, the main route as centralizing axis of the fabric is mortified in its centrality: due to the great distance between the two opposite sides of commercial unit fronts[15] and due to the presence of motor vehicle traffic, the possibility of communication and pedestrian movements from one commercial front to the other one, together with comparative shopping, decreases strongly. Furthermore, the user who has not yet parked his vehicle, given the speed of travel, is unable to perceive the centralized route as the heart of the fabric's commercial life. The secondary façade, the rear, spontaneously becomes a contradictory and incongruous main front where each store unit is accessed through the service door, crossing the goods loading/unloading flux and other technical functions such as the transport of waste.

A similar situation can be found in the case of the *Broadway-Crenshaw center + May company store*, Los Angeles (1954). Also, this case shows a specialized fabric which is still, problematically, crossed and centralized by a driveway. It is also possible to find a greater hierarchical differentiation within the component building

---

[15]Distance reminiscent of the dimensional transformation of the route at the time of its specialization, when its width is increased to accommodate a greater vehicular flow, that of the mass use automobile.

units group: to a first structure characterized by the presence of a special commercial unit, the department store with the function of attracting the pedestrian/customers flow a second polarity (*May company store*) is added as a second anchor, according to a spontaneous process, thus defining an *ante litteram* two anchors "dumb-bell" scheme. Except from a peculiar and distinct development, the *Broadway-Crenshaw* case presents analogous problems with *Westchester business center* case: they both well represent this intermediate phase of fabric's specialization process, full of contradictions and uncertainties.

3.2.3 Mature Phase: the Shopping Center as Specialized Bypassed Nodal Fabric

If in the historical centers/downtowns routes, specialization and bypass formation process leads to the so-called *pedestrian malls*, in the suburban territory these same processes will lead, instead, to the formation of the mature commercial specialized suburban fabric, the typical *mall*, that after being formed and concretized into a mature type during 1960s decade, will then be used for at least the next two decades and subsequently, with some critical revisitation, up to present day.

This phase of the formative process becomes clear by looking at some case studies, such as the 1942 *Community center* designed by E. Saarinen, in Willow Run—Michigan, where the specialized fabric is centered on an exclusively pedestrian route. The pedestrian axis is characterized by an organization in section that comprises a linear garden that divides it into two, and, in plan, by a double polarization with commercial units—*anchors* positioned at the ends of the pedestrian route that hierarchize all the fabric.

It is only in Victor Gruen's works[16] that the spontaneous formative dynamics of the commercial suburban fabric find a mature critical design interpretation. Only with and after his work the critical interpretations advanced up on several fronts and by several voices, finding a stable and lasting conformation that will be applied and repeated, with necessary variations and personal interpretations (as if they were part of the spontaneous heritage) therefore becoming type (on type notion: Caniggia and Maffei 1979, 110–122).

The interpretation of the fabric's formative process proposed by Gruen starts from the changes and transformations that involved the base commercial building unit, in its projects developed in his native Austria and then in America once he had migrated. At the core of these works there was, above all, the problem of integration between urban fabric life and commercial unit, questioned by studying the interface/display window conformation and by developing the design of commercial unit as an indoor

---

[16] Victor Gruen (1903–1980), American Austrian architect, specialized in the design of commercial spaces and suburban shopping malls and in the last part of his career urban fabrics with commercial specialization through various *pedestrian mall* projects for American and European cities (Gruen 1960, 1965, 1973; Wall 2005).

micro-urban fabric, without large divisions between the areas reserved for sellers and those for customers (Wall 2005, 18–48). All these works characterize a first phase of the critical reflection by Gruen on a spontaneous process,[17] a reflection that will be extended, later, also to urban fabric scale.

The design proposal for the unbuilt *Shopping Center for 194X*, Syracuse-NY, 1943, is particularly eloquent as regards the distinction between fabric's specialized elements: parking lot, centralizing route space, commercial units aggregated according to a defined hierarchical order, in which the presence of special units is clearly recognizable.

In this project, moreover, the centralizing axis, almost areal (rather than linear) for its extended dimensions, moves inside the subdivision block, thus using the suburban subdivision routes' network as a bypass route.[18] Similarly, considerations can be made for other contemporary projects such as the one for the *Linda vista shopping center*, San Diego-California, 1944. Here, as in Gruen's project, the driveway bypass route forms a ring around the specialized commercial (and pedestrianized) fabric; between bypass ring route and commercial fabric, parking areas are located that allow the travel modalities' exchange, from car to pedestrian, and vice versa.

With *Northland center* project in Detroit, and then for the *Southdale center* in Edina-Minnesota, the reflections and experimentations on the structure of the suburban commercial fabric take on a stable form. The former case still faces a fabric polarized by only one anchor, whereas the canonic and typical "*dumb-bell*" scheme will see the presence of two anchors. This project is greatly interesting because of the relationship existing among the special fabric and the territorial routes: the fabric is served by a main bypass route that splits from the main territorial motorway route network, then serving, through other secondary rings and radial routes, a vast ring-shaped parking area for intermodal exchange. The parking area is shaped as a ring, thus isolating the commercial fabric from the territorial routes and the territory itself.

With the *Southdale center* we meet the consecration of the *dumb-bell* scheme, a fabric polarized by two anchors positioned at a distance of 200 m at the opposite ends of a centralized, pedestrian route. This route in this project is still characterized by large width in section, but it will gradually shrink in the future developments to regain the dimensions of a traditional pedestrian urban route, without the vehicular specialization and use. Another innovative and very important aspect of *Southdale* is the fact that it is "fully enclosed". The harsh climate of Minnesota and the new achieved technical possibility of air-conditioning large areas led to a complete coverage and closure of the centralized route, which until then was open-air with canopies to unify commercial units' façades, to shade the showcase windows and to guarantee a certain continuity and a covered route along the commercial units fronts. Total coverage and air-conditioning bring an important transformation

---

[17]That process that, with the self-service system introduction, generated special commercial units, such as *department stores* and, later, *supermarkets*.

[18]The localization of the centralizing axis of the specialized fabric within the block, or in any case in an area that acts as antipolar in relation to the routes network, anticipates by a few decades the theory of commercial special fabric design described in: Maitland (1985), 162–171.

in the suburban commercial fabric. It is an operation comparable to a *fabric scale knotting* (definition of "knotting" Strappa 2010—http://www.wikitecnica.com/ann odamento/—consulted on 21/08/2019) that definitely declares the pedestrian and commercial specialization of the centralizing route and surrounding *contrada*.

The knotted centralizing route, which subsequently becomes the *contrada's* special "room", also brings the substitution of the principle of aggregation of units (contradicted by the internal dynamics of distribution of the commercial units dominated by continuous re-division) by the subdivision (Strappa and Buongiorno 2019). Taking into account the facilitation due to the use of a punctual constructive system of concrete/steel pillars and beams, in the synchronically knotted fabric it is more easily possible to make a distinction between component commercial units by looking at the distribution system than at the constructive one. The subdivisions are progressively lighter and lighter in the process, the façades of each commercial unit with the traditional display/showcase space and access door lose their reason of existence, turning into mobile or extremely permeable elements, physical or also just visual.[19]

The original critical reflection, by Gruen, on commercial unit design conceived as micro-fabric (obtained by transforming and opening its façade into an exhibition area for new users increasingly accustomed to an active self-service role in purchasing) is brought in *Southdale center case* at a broader level, by involving the whole commercial aggregate, reconnecting the specialized commercial fabric with the internal commercial units' micro-fabrics, so establishing a richer and more complex hierarchical order.

3.3 Building Scale

Substratum—row houses (shop house synchronic variant) and their aggregation on the centralizing axis.

The traditional workshop/store, which occupies the ground floor in one of the row house type synchronic variants, represents in itself the arrival point of a long formative process. It started from the spontaneous trade practice on the road and progressively stabilizes until assuming the characters of a modern store.

This store is, first and foremost, a space for a business that sees the seller as a very active figure at the service of the customer, which instead plays a more passive role

---

[19]The gradual disappearance and permeabilization of the commercial façade is the result of a long process, clearly described in Walter Benjamin's descriptions of Parisian Passages:

> …può mutare l'intera Parigi in un *intérieur,* in un'abitazione le cui stanze, non divise da soglie come le camere vere e proprie, sono i quartieri, così d'altro canto, la città può schiudersi al passante da ogni parte come un paesaggio senza soglie.

> … the whole of Paris can transform into an *intérieur,* into a house whose rooms/urban districts are not divided by thresholds like the actual rooms; so the city can open itself to passers-by on all sides like a landscape without thresholds (Benjamin 1986, 551).

(these conditions will change only with the contemporary customer, involved in the purchase operation substantially through the *self-service* system). The space of the modern Roman workshop/store, belonging to the traditional pre-industrial fabric, is not very different from the workshop of the ancient Roman fabric (Giannini 1970). In the latter, the interface barrier between customer and seller is located almost at the border of the commercial unit, with very large openings for exchange and communication. Thus, the sales desk/interface node, during the commercial activity hours, in which the doors and closures on the road are open, appears to be placed almost on the road, to intercept the urban fabric's life directly, without mediation.

The commercial space of the modern row house, heir to the ancient Roman workshop, is the result of a (first) specializing action. This specialization process transforms the store's spatial organization by interjecting inside a part of the urban fabric, creating a clear division within the interior space, between space occupied by the customers, communicating with the street through the physical passage of the entrance door, and space for the sellers, less and less in contact with the street, but in touch with the customer's new "semi-public space" inside the commercial unit, readable as an indoor extension of the public space of the street.

This process of interjecting experiences an acceleration because of the combined action of technical progress in the field of electricity and production of glass sheets for large openings and windows on the road. Together, these factors contribute to the formation of the *boutique*. *Boutique* adds the presence of the showcase to the features already present. The introduction of the showcase window is closely linked to the technical innovations in the productive process of large glass sheets in France in the seventeenth and eighteenth centuries (Saint-Gobain glass factory), which can be built with a reduced presence of metal frames, thus not disturbing the customer's view; at the same time, the innovation of showcase windows is strictly linked to the introduction of a new electric lighting system that allows to illuminate the goods from the store's interiors, with a reduced fire risk and with the possibility of directing the lights so as not to dazzle the buyer. The communication with the urban fabric becomes no longer only physical and tactile but also visual: large showcase openings make it possible to view the range of products on sale, thus enabling the customer to play an active and independent role in the sale/purchase process, choosing in the spectacular succession of shop windows.

Examples of this *boutique* can be found within the Parisian *passages* specialized aggregates, characterized by the sparkle of precious goods, spectacularly enhanced by the natural zenith light that descends from the transparent route covering, combined with the directed artificial electric lighting.

The *boutique* will have important developments in the American area, with the first commercial unit projects of Victor Gruen, in which the urban route is welcomed inside the display case that becomes concave and linearly more extensive (Wall 2005, 18–48).

3.3.1 Boutique Specialization and Special Commercial Unit Formation

This phase sees, in the urban area, the re-melting/re-subdivision and specialization of the single-base commercial serial units. It opens perspectives that span from types such as the covered market building, going through *passages* and *galleries*, to reaching the *department stores* and *grand magasins*. The specialization takes place through the formation of an interior fabric inside the aggregated building unit/block, characterized by a specialized internal main route, in continuity with the external urban fabric's routes.

The commercial unit's transformation and specialization follows a process similar to that of other special building types, such as the *Palazzo* or the *Monastery* (Strappa 1995, 129–136). As in the latter, in the formation of the modern special commercial unit[20] the scale of the single-base serial building unit enters in close relationship with that of the urban fabric, one integrating the other.

The formation of special commercial unit—from the specialist building of the covered market to the formation of the department stores, passing through *passages* and *galleries*—takes place at the aggregate scale, within the fabric, through the small base units' recasting and by overturning/interjecting the urban fabric's routes (or portions of those) inside the urban block, finally knotting them in the creation of a special nodal room/compartment.

In projects such as that for *Cour Batave* or *Grands Magasins du Printemps*, both located in Paris, the derivation from the basic block is very evident, transformed by recasting previous units and by specializing the block's courtyard/rear.

In the first case, *Cour Batave,* reading the recast aggregation process is immediate and quite evident. In the second case, *Grands Magasins,* on the one hand the internal organization by aggregation of cells and specialization of the aggregative/architectural organism there is a knotting of routes into a central multiple heights special nodal room/compartment, while on the other (also because of the use of a punctual structural system that allows extreme flexibility and independence between structure and distribution), due to the commercial fabric's synchronic formation the whole system can be seen more in terms of subdivision rather than aggregation. The spatial subdivision, however, follows anthropic spatial rules, analogous to those characterizing the urban fabric aggregative dynamics, as demonstrated by the use of elementary square cells of 6/7 m. The subdivision principle, in the formative and historical process, will regulate the formation of urban and suburban commercial specialized fabric up to the present (Strappa and Buongiorno 2019).

Although the analogy with the *Palazzo* and *Monastery's* formative/transformative process and some of the already mentioned case studies suggest an aggregation and

---

[20]This book refers to department stores, grand magasins, as special commercial units. These special buildings are studied with more attention than, for example, the market type, since in them the process of derivation from the urban fabric is more immediately recognizable. In the case of the market building, instead, the special building type acquired soon a very high autonomy level.

specialization of the commercial units forming the block with subsequent formation of a special knotted space, it often happens that the aggregation/subdivision
of/into cells organized within an urban route interjected in the block does not give
birth to a multiple heights special compartment, as for example in the majority of
American department stores (Longstreth 2010). This does not diminish the urban
fabric character and behavior of these structures, confirmed, instead, by the strong
continuity between the internal commercial fabric and the external urban fabric. In
this regard, it is sufficient to refer to historic New York department stores such as
Macy's, *Bloomingdale's* or *Saks Fifth Avenue*. Walking through these commercial
spaces, it is possible to notice that, since the internal routes are in strong continuity
with the external urban ones, the barrier between outside and inside becomes weak
and minimal.

The transformation and recasting process for the formation of the special store
unit proceeds in two distinct directions:

– recasting of the base commercial units of an entire urban block or just a portion,
  and formation of a special commercial unit with or without a special knotted nodal
  space (e.g.: department stores, covered markets, department stores);
– recasting of the base commercial units belonging to distinct blocks facing the same
  urban route, with the formation of a special compartment/room on an existing
  or new route, by knotting routes and possibly covering them (e.g.: *passages*,[21]
  galleries, shopping arcades).

These two development directions, characterizing the transformative/specializing
process are distinct and complementary, both at the building scale and to the
aggregative one. Today's malls spaces show the coexistence of both transformative
modes.

3.3.2 Further Specialization of Base, Aggregated and Special Commercial Units: Supermarket and Self-Service Fabric Organization

If the formation of special commercial units was generated by a block's specializing
process (a fabric interjected in a building), similarly to the specializations that had
already taken place in history (e.g.: *Monasteries, Palazzo*, etc.), with the *self-service*
sale/purchase system advent, a strong acceleration of the process appears on the
scene. It has a strong impact not only on the large commercial units discussed in
the previous paragraph, but also on the smaller ones, such as base serial commercial
units and medium-aggregated ones.

The *self-service* system diffusion in the first experimental projects, those for
medium-aggregated building units such as the first *supermarkets*, strongly catalyzes
the process of opening and connecting the unit store toward the urban fabric.

---

[21]On Parisian *passage* (Geist 1995; Lemoine 1989; Benjamin 1986).

An interesting experimental phase during the first decades of the twentieth century will lead to projects in which the absorption of the urban fabric within the commercial unit and vice versa is very clear. The *supermarket* (Nencini 2017), created to increase consumption in an America exhausted by the effects of the 1929 crisis, owes its existence and fortune to the development of the *self-service* sale/purchase system. The large assortment, in quantitative and qualitative terms, of goods for sale, combined with the quantitative decrease of qualified workers need for sales/purchase operations, allowed at the time of its debut significantly lower prices for the consumer and thus it outperformed the competition of traditional businesses.

In its formative process the *supermarket* results from a series of experiments carried out in the *drive-in market* (Longstreth 2000).

The link between this new type of commercial unit and vehicular traffic is very strong: sale/purchase activity of larger quantities of food (and not only food) is made possible by a low price business policy, and also above all, by the progressively growing use of cars (which in the succession of produced models becomes more and more voluminous and capacious) for the transport home of the purchased goods. A greater volume of purchases is a consequence, and at the same time constitutes a contributory cause, for the totally new great distances imposed by the suburban settlement model. Within the new suburban structure it is not convenient and possible to travel daily, or very often, from the residence place to the special commercial unit/supermarket for purchases. A very important contribution to the supermarket's success and to the new qualitative and quantitative sales trends was also the introduction of large-scale use of domestic technical tools and accessories, such as the refrigerator.[22]

The contemporary action of complex factors leads to the formation of a special commercial unit dedicated to the food commerce (later the range of goods for sale would have been extended). This commercial unit, in its first moves, shows precisely its derivation from the interjecting of the fabric and suburban routes within the basic commercial unit.

The *Clarence Saunders "Self-serving store", 1917* (Liebs 1995, 120–121), as one of the first cases of application of the self-service system, is structured in an exemplar manner, on a single specialized route marked by the entrance/cash/turnpike polarity/nodality and by intermediate nodes, positioned at the bottom of the commercial unit, more or less in the middle of the route journey, devoted to special goods sale (fresh and perishable goods, meat, fish, etc.). This specialized route is comparable to the main three-dimensional specialized route of the department store, but at the same time it constitutes its extreme evolution: it is an obligated one-way loop journey route, that exposes customers, forcedly, to the whole offer of goods for sale and is, at least initially, the only existing route usable by customers.[23]

---

[22]On development and spreading of *supermarket* type in America and Europe: (Scarpellini 2001; Scarpellini 2006; Scarpellini 2008; Liebs 1995, 17–137).

[23]The use of loop one-way routes, which are also specialized as regards the direction of fruition, has a history rich in mutual exchanges between two worlds: commerce and museums. On the one hand, there are examples of museums with loop routes such as the Guggenheim museum in New York

The introduced route shows a very strong continuity with the surrounding urban fabric routes' network, up to the point of experimenting its vehicular drive-through possibilities, in an extremely critical and creative interpretation of the *drive-in market*, the *Automarket*, built during the 1920s in the United States, in Louisville—Kentucky (Liebs 1995, 122). In this particular supermarket, customers do not leave their car to make their purchases. The special commercial fabric settles on a hyper-specialized vehicular route. The several exhibition units settle on the route's pertinent strips; the route is polarized by the access turnpike and by the payment node.

To summarize, while the *Department store*, as well as the *Passages* and the *Galleries* marked the building unit's specialization and the accentuation of a urban fabric logic of organization of the interior space, the *supermarket* has led to further results in the route's specialization, by applying the *self-service* system to an interior fabric dominated by a loop route.

The current phase of the process sees important changes regarding the entrance/exit, which progressively tend to coincide, becoming much less cumbersome. Specifically, the advent of very fast digital payment methods makes possible to thin more and more the barrier of the cash desk/payment node, with the inclusion of fast and self-service or remote digital payment machines, as in the avantgarde AmazonGo store, Seattle, USA (AmazonGo, https://en.wikipedia.org/wiki/Amazon_Go—consulted on 16/09/2019).

3.4   Graphic Schemes: Urban and Suburban Commercial Special Fabric Formative/Transformative Process, with Examples

The schematization of special commercial fabrics' formative and transformative processes that follow constitutes supporting material developed *in itinere* during the investigation work and is proposed here as a synthesis tool.

The identified urban and suburban context's process phases should be read as indepth analysis and illustration of the macro-phases identified in this Chapter. In some cases the macro-phases coincide with the schematization phases, while in other cases the scheme illustrates different sub-phases included in the macro-phase described in the chapters.

---

and the Vatican Museums in Rome, on the other, cases of commercial spaces such as that of the first supermarkets or the more recent *Ikea* special commercial units. For the relationships between museum space and commercial space: (Benjamin 1986, 542; Teufel and Zimmermann 2015).

3.4.1 Formative/Transformative Process in Urban Context

As regards the process in an urban context (traditional urban fabric: historical centers and downtown), the starting point of the schematization by phases is a synthesis of basic fabric's formative process, organized with a *contrada* specialization (Fig. 3.1).

In the following phases, the typical basic blocks' specialization dynamics operate on the basic fabric common processes in the formation of different types with different outcomes of special serial and polar construction.

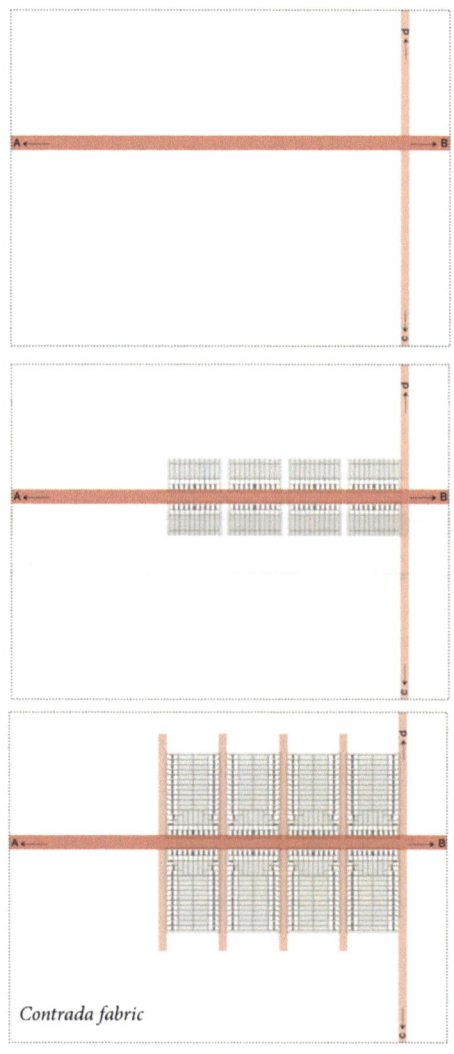

Fig. 3.1 Base "contrada" fabric formation *Source* V. Buongiorno

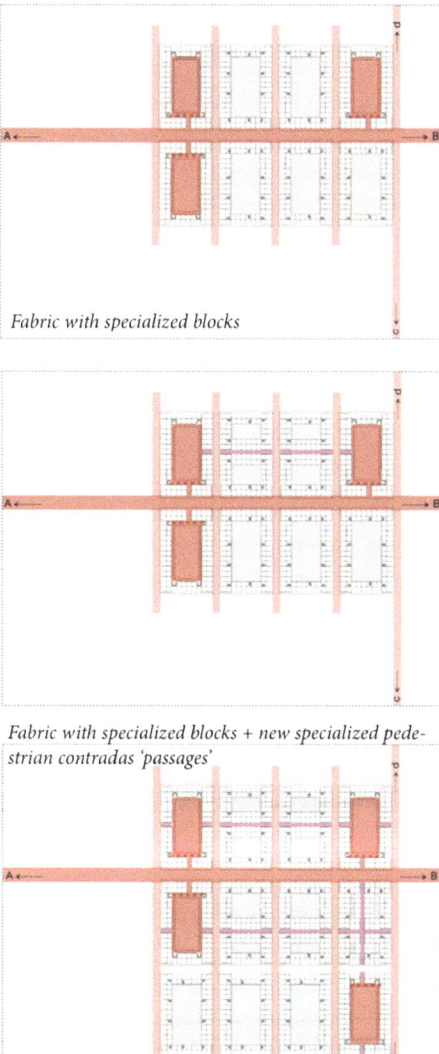

Fabric with specialized blocks

Fabric with specialized blocks + new specialized pede-strian contradas 'passages'

Fig. 3.2 Transformation and special commercial fabric formation *Source* V. Buongiorno

The basic urban fabric's routes overturning within the block, forming a real fabric inside the new specialized block, is marked by the settlement of these new overturned routes polarized by vertical ascent routes (main and secondaries stairs), in order to make communication possible between the various arrays components to the various planes.

The example of department stores, both in Paris and in American department stores, well represents this phase of the fabric specialization process.

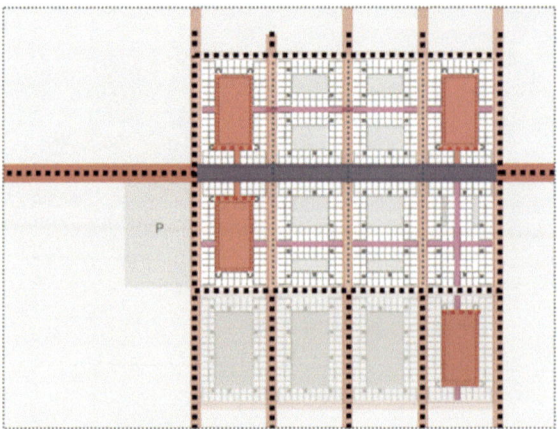

Specialization of the whole fabric: pedestrianization and knotting with route's coverage

Fig. 3.3 Transformation and special commercial fabric formation *Source* V. Buongiorno

This is followed by the formation of a network of autonomous and specialized trade routes which connect the blocks with commercial specialization and which are polarized by them. This network is necessary and organic to a system now characterized by a significant level of use specialization. The autonomous and super-imposed network of specialized routes allows bypassing the paths of the basic traditional urban fabric on which the transformation process has been set, which, with the specialized use, automotive for example, gradually ceases their capacity to host a mixed specialization (Fig. 3.2).

An example of this phase is the case of the Parisian passages that connect specialized poles of the city and allow the buyers to avoid, therefore bypass, the great axes of driveway and to do "comparative shopping" along the new dedicated routes.

Other examples of this phase are the so-called "parallel cities" specialized fabrics, underground or at high altitude, such as the North American underground cities and the American or Asian skywalks.

The "final" phase of formation/transformation of the fabric with commercial specialization in the urban context sees the definition of a whole portion of fabric dedicated to the specialized function (Fig. 3.3). This is accomplished through the delimitation of this portion and the definition of limitations of the types of journeys allowed inside it, with the creation of areas with limited traffic and pedestrian areas, for example. This phase of the process projects the specialized fabric into a vast urban and territorial dimension, transforming it. The area of specialized fabric takes on the role of specialized nodality of the network of territorial and urban routes within the territorial and urban organism. The bypass route that delimits it is a specialization of the same network of routes in order to serve the portion of fabric and make it reachable on a large scale.

For this phase of the fabric the following examples might be considered: the experiences of the pedestrian malls; portions of fabric with lower levels or others of specialization, not completely pedestrian, or completely pedestrian and knotted covered routes.

3.4.2 Formative/Transformative Process in Suburban Context

As regards the formation/transformation process of the commercial specialized fabric in a suburban context, starting from the formative macro-phase of fabric settlement on the dividing lines, followed by the specialization with bypass, up to the mature type of bypassed special fabric, we propose here a schematization of the process, decidedly less diachronic, related to the last phase that is assuming the adoption of the mature type of the suburban mall (Figs. 3.4, 3.5, 3.6 and 3.7).

In the sequence of schemes, the sub-phases of training are proposed within today's macro-phase of the mature type mall.

Starting from a basic scheme already characterized by the presence of an inter-modal parking/exchange area of two special commercial units, A1 and A2, as in a specialized strip scheme and off-center in relation to the territorial route/dividing line—on which the real strip, instead, is settled—we see in the first three phases the formation of a fabric on the main commercial route PP pertinent strips, which is polarized by A1 and A2 and constitutes a centralizing axis. This "base" fabric of serial units, polarized by A1 and A2, is characterized by dynamics of the recasting of serial units for the formation of intermediate units and by the presence of synchronic corner variants, with smaller units (Fig. 3.4).

The extension of the fabric takes place, in the process as in the reality of contemporary malls, with the establishment of new paths polarized by new polar commercial units.

These paths are the continuation of the main commercial route. The P.P. extension continues, in the typical case, until it reaches the conformation of the polarized circuit/loop path by a minimum of 3 or 4 special commercial units-anchors. Even in the complete conformation the formed fabric appears as a "basic" fabric: although it is used for a specialized function, in its morphological structure it presents the characteristics of the basic fabric. It is characterized by a minimum level of specialization, that of the *contrada*, in which the block, even if complete, does not exist as an autonomous entity but lives in complete subordination to the route(s) from which it generated (Fig. 3.5).

In the case of complete blocks, the pertinent courtyard area—similar to the internal courtyard subdivided into areas pertaining to the individual units of the traditional European fabric's base block—is occupied by the technical equipment of the individual serial commercial units and by the pertinent technical route.

To illustrate this phase and sub-phases of the process, consider the cases of typical western suburban malls.

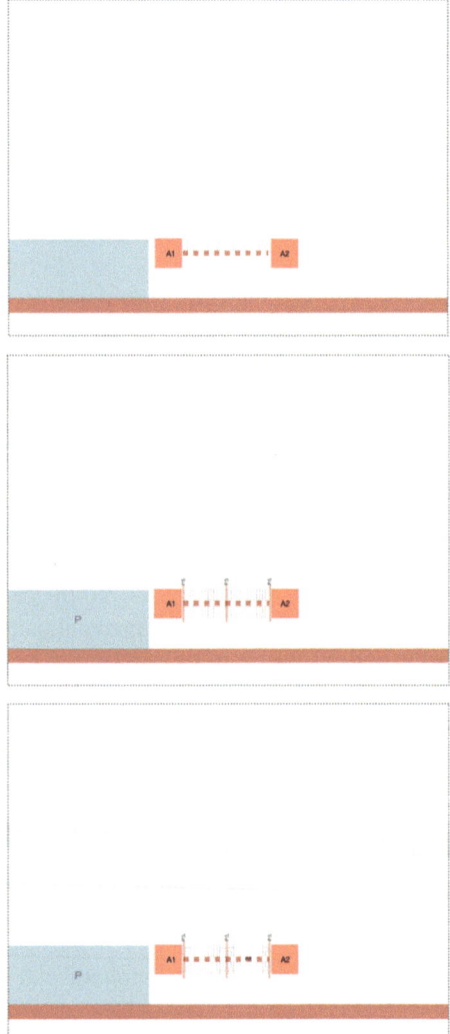

Fig. 3.4 Base "contrada" fabric formation *Source* V. Buongiorno

The most recent phase of the process sees the transformation of the block with the specialization and knotting, and with the formation of a special non-commercial unit in the block's special compartment (Figs. 3.6 and 3.7).

This phase partially coincides with the future transformation perspective outlined in Chap. 4, although in the already existing cases of malls in which this dynamic is observable, the specialization of the block must deal with an abstract character that is unique, and in which super-dimensioning and gigantism isolate the specialized fabric from the urban/territorial organism in which it is inserted.

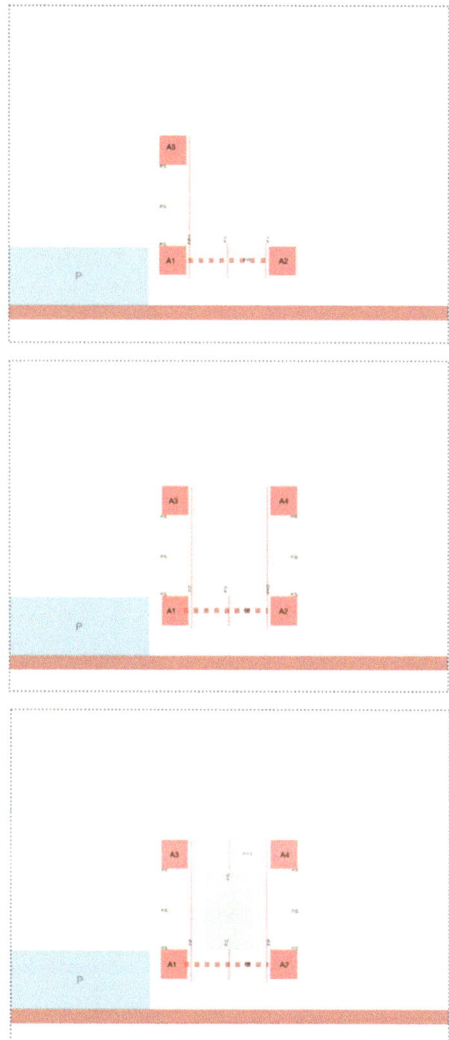

Fig. 3.5 Base "contrada" fabric formation *Source* V. Buongiorno

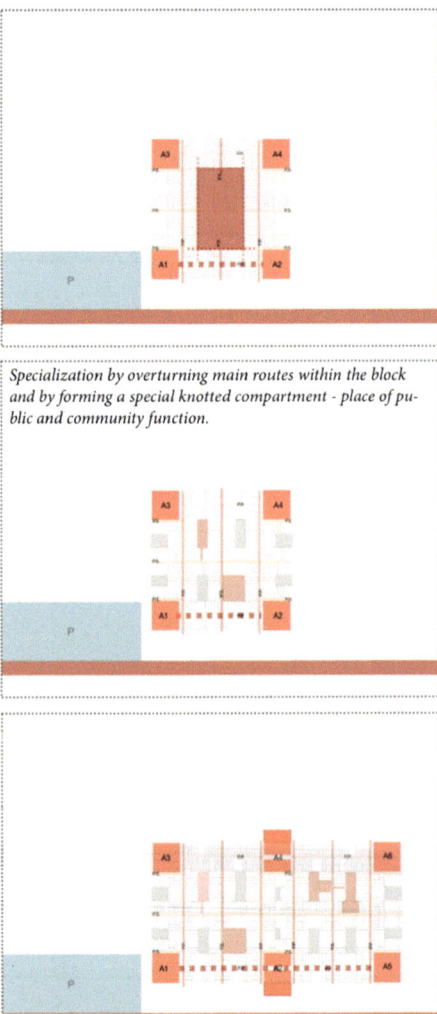

Fig. 3.6 Transformation *Source* V. Buongiorno

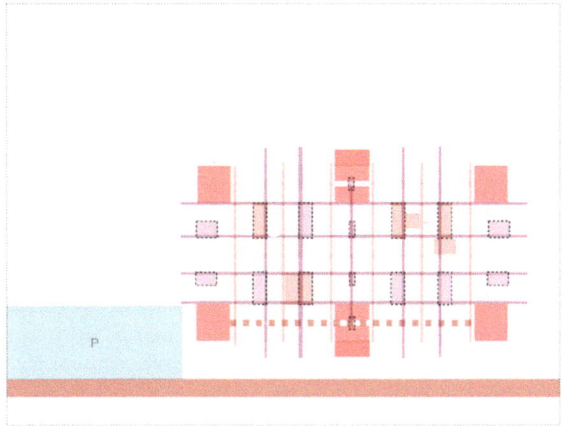

Not strictly commercial (public and community) special routes' network superimposi-
tion, polarized by the special public community knotted compartments.
Routes, like new 'passages', new contradas, bypass pedestrian commercial traffic. The
routes' network extends horizontally and vertically, having special knotted compart-
ments as nodes and pivots.

Fig. 3.7 Transformation *Source* V. Buongiorno

References

Anderson S (1986) On street. MIT Press, Cambridge-MA

Benjamin W (1986) Parigi, capitale del XIX secolo. I "Passages" di Parigi. Giulio Einaudi editore, Torino

Caniggia G, Maffei GL (1979) Composizione architettonica e tipologia edilizia: 1. Lettura dell'edilizia di base. Marsilio, Venezia

Case Sheer B, Stanilov K (2004) Suburban form: an international perspective. Routledge, New York/London

Chung CJ, Inaba J, Koolhaas R, Leong ST (2001) The Harvard design school guide to shopping/Harvard design school project on the city 2. Taschen, Koln

Corbusier LE, Jeanneret-Gris CE (1929) La Rue: texte manuscrit pour l'Intransigeant 05/1929. Fondation Le Corbusier Archives, Paris

Dunham-Jones E, Williamson J (2011) Retrofitting suburbia. Urban design solutions for redesigning suburbs. John Wiley & sons, Hoboken-NJ

Fariello F (1963) Architettura delle strade. La strada come opera d'arte. Tipografia della Pace, Roma

Giannini S (1970) Ostia. in Quaderni dell'Istituto di elementi di architettura e rilievo dei monumenti, n. 4. Università degli studi di Genova, Genova

Gruen V, Smith L (1960) Shopping towns USA: The planning of shopping centers. Reinhold, New York

Gruen V (1965) The heart of our cities. The urban crisis: diagnosis and cure. Thames and Hudson, London

Gruen V (1973) Centers for the urban environment: survival of the cities. Van Nostrand Reinhold, New York

Hindley G (1972) A history of roads. The Citadel Press, Secaucus-NJ

Lacroix D (2003) Morphogenèse de Sainte-foy: le dilemme de la banlieue moderne entre structures héritées et forme urbaine nouvelle une analyse morphologique. Ulaval, Quebec

Liebs CH (1995) Main street to Miracle mile. American roadside architecture. The John Hopkins University Press, Baltimore/London

Longstreth R (1997) City center to regional mall. Architecture, the automobile, and retailing in Los Angeles, 1920–1950. The MIT press, Cambridge-MA

Longstreth R (2000a) The Drive-In, the supermarket, and the transformation of commercial space in Los Angeles, 1914-1941. The MIT press, Cambridge-MA

Longstreth R (2000b) The buildings of main street. A guide to american commercial architecture. Altamira press, New York/Oxford

Longstreth R (2010) The American department store transformed. 1920–1960. Yale University Press, New Haven & London

Maitland B (1985) Shopping malls: planning and design. Construction press, London

Moraglio M (2007) Storia delle prime autostrade italiane (1922–1943). Modernizzazione, affari e propaganda. Trauben, Torino

Nencini D (2017) "Supermercato". In http://www.wikitecnica.com/supermercato/-consulted on 21/09/2019

Rudofsky B (1969) Streets for people. A Primer for Americans. Doubleday & Company Inc., New York

Rykwert J (1982) La strada: utilità della sua storia. In: Anderson S (ed) *Strade.* Dedalo, Bari

Rykwert J (1995) Imparare dalla Strada. In: Rykwert J (ed) Necessità dell'artificio. Arnoldo Mondadori, Milano

Scarpellini E (2001) Comprare all'americana. Le origini della rivoluzione commerciale in Italia 1945–1971. Il Mulino, Bologna

Scarpellini E (2006) Esselunga. Agli albori del commercio moderno. Art, Bologna

Scarpellini E (2008) L'Italia dei consumi. Dalla Belle Époque al Nuovo Millennio. editori Laterza, Bari/Roma

Strappa G (1995) Unità dell'organismo architettonico. Note sulla formazione e trasformazione dei caratteri degli edifici. Edizioni Dedalo, Bari

Strappa G (2010) "Hnotting". In http://www.wikitecnica.com/annodamento/-consulted on 21/08/2019

Strappa G (2016) Architecture of the territory. In: Strappa G, Carlotti P, Camiz A (eds) Urban morphology and historical fabrics. Contemporary design of small towns in Latium. Gangemi Editore, Roma

Strappa G, Buongiorno V (2019) Commercial Urban fabrics updating. Retail globalization and shopping cultural areas. In: Charalambous N, Zafer Cömert N, Hoşkara Ş (2019) CyNUM 2018 conference Urban Morphology in South-Eastern Mediterranean Cities: challenges and opportunities Proceedings. Cyprus Network of Urban morphology, Nicosia-CY

Teufel P, Zimmermann R (2015) Holistic retail design. Reshaping shopping for the digital era. Frame Publishers, Amsterdam

Vachon L, Luka N, Lacroix D (2004) Complexity and contradiction in the aging early postwar suburbs of Québec City. In: Case Sheer B, Stanilov K (2004) Suburban form: an international perspective. Routledge, New York/London

Wall A (2005) Victor Gruen: from urban shop to new city. Actar, Bacelona

Chapter 4
Retail Spaces Crisis and Future Transformative Process

The texts useful for us, the descriptions, the maps, the cadastral maps, are not objective documents. Every reading is, inevitably, a critical interpretation. We read things with our eyes and elaborate images with the mechanisms that our mind, unique and unrepeatable, uses in design: at the end we find, however many objectivity efforts can be made, what we are looking for (Strappa 2012, 19).

Further specialization and/or de-specialization.

The formative and transformative process described so far led to the mature form of the commercial suburban fabric type. A long and complex process characterized for its specialized character and for realizing the conditions for a further future specialization of use of the aggregates/fabrics.

As seen in the previous chapters, the fabric specialization, in the suburban context, as well as in the traditional urban context, is the cause and, at the same time, consequence of the transformation of the fabric, or a portion of it, into a territorial node. This consecration of the fabric to the role of territorial node is implemented through a morphological base action: the division and the specialization of the traffic/exchange flows (divided by speed, means, direction) made possible, thanks to the realization of special bypass routes.

In the current condition, after the advent and application of the route's specialization to the retail world—as well as to a very wide range of areas—routes for virtual flows of data and information are added to the complex flows and physical (automotive and not) traffic of people and goods resulting from technical evolutions occurred throughout the twentieth century.

The development of *information technology* contributes to the further specialization process that generates new very special routes, able to bypass a great portion of congested physical routes, by using digital channels virtually accessible through domestic terminals and personal devices.

The advent of *e-commerce*, in this context, can be read as a further specialization of the goods purchasing and consumption dynamics. Digital technology tools, as a virtual multidimensional bypass ring route, allow us to avoid part of the physical route

V. Buongiorno, *Suburban Retail Spaces*,
SpringerBriefs in Geography,
https://doi.org/10.1007/978-3-030-54991-6_4

and the related distance to go to the store for goods' purchasing. The shopping activity is offered within the virtual environment, and only the goods' delivery will travel on the traditional physical routes,[1] even though organized within a new distribution and logistic system, that by bypassing the physical retailers connect directly from the territorial scale warehouses to the individual consumers.

If the new information and goods' routes make a new bypass, it is not yet possible to do so for all product/merchandise sectors, and even in the most digitized sectors, the dynamic is very far from being simple and linear. On the one hand, it is possible to observe a progressive virtualization of the sale/purchase operations, especially for current and daily use goods, as food for example. On the other hand, for example as regards high-end goods, characterized by a significant qualitative added value, the e-commerce virtual purchase/sale dynamic proves to be still insufficient.

The need for an overall product experience to finalize its purchase raises the question of the need for the *mortar&brick* store.[2] While the physical store seems destined to disappear, substituted by its virtual counterpart, it becomes clear that an irreplaceable element of the *in presence* buying dynamic, the sensorial/experiential and cultural aspects specific of the *mortar&brick* store, are not simulable and/or achievable in a virtual environment.

The complex questions synthetically outlined constitute a base and a substratum, for a future transformative process for commercial environments in general and, specifically, for suburban shopping malls.

The transformative process outlined in the following paragraphs is intended as a possible future phase, a critical continuation of the formative and transformative process read and interpreted in the current and past built environment of commercial suburban fabrics at the various scales and in the specific scientific literature (Caniggia and Maffei 1987).

4.1 Territorial and Urban Scale

At the territorial scale the introduction of *e-commerce* has caused important changes in the delicate balance between goods' distribution system and retail sales, a relationship in which the former tends progressively to replace the latter.

There is a horizon, futuristic but in part already present, in which physical sales spaces tend to decrease in quantitative terms, to focus energies and efforts on qualitative values. Furthermore, the presence of large storage areas for goods and reserves for retail sale doesn't make sense anymore, because very often the physical store

---

[1]When goods are (still) physical. Consider, in fact, the sectors where digitization has impacted so strongly as to minimize even the physicality of the goods, now available in a virtual format, as in the cases of 'publishing, bookshops, newspapers, music stores, etc.;.

[2]ATKearney agency's Report: The future of shopping centers (https://www.atkearney.com/doc uments/20152/986752/The+Future+of+Shopping+Centers.pdf/6455ae6f-f430-2fe7-2856-ef6711 53d29a—consulted on 21/04/2019).

finds greater profit in concentrating its energies on the overall sensorial and cultural shopping experience than on the quantitative offer and distribution of goods.[3]

In this situation, a role of increased relevance is played by the distributors' territorial warehouses that sell and deliver goods to both physical stores and individual e-commerce customers.

These distribution hubs, located in territorial nodal/polar positions, tend to progressively replace the shopping mall in its, albeit partial, function of goods' exchange and distribution node. The distribution chain now tends to exhaust its process within the distribution hub, from where the goods, ordered and paid for online, are sent via postal/express courier service, through the network of specialized motorway routes (and probably in a near future via drone using a three-dimensional route network[4]).

Especially in suburban contexts, the mall, with its physical commercial units and its internal hierarchical balance, ceases to be the node of the goods exchange/distribution. It passes this role to the large distribution hubs, not accessible to the single users but extremely important for them, since the distance that separates them determines the waiting times for the delivery of goods purchased online.

In the perspective of continuing the formative/transformative process read and interpreted in the current built environment, it is possible to glimpse a hypothetical new future phase, able to rebalance weights on old and new nodes.

A series of transformative interventions and updating of the commercial territorial organization outline the new critical hypothetical phase:

– updating and transformation of the territorial routes network through the further specialization of parts of the existing network and/or through construction of new integrative/complementary parts in order to allow for the exclusive special use for the purposes of commercial travel.[5] The preparation of dedicated routes, as well as for the future use of drones for deliveries of goods purchased online, could also take place by transforming the existing vehicular, motorway, cycle and water lines' network;
– transformation of the territorial nodes (some of the many disused ex-regional shopping malls) in new distribution hubs, according to a process of recasting;
– transformation of territorial nodes, into nodes with multiple specializations, according to a process of *de-quantification* and *de-specialization*;

---

[3]On sensoriality and commerce: White (1991).

[4]See the experimental distribution system Amazon Air implemented by Amazon which includes the construction of a private airport for goods fast transport and a fleet of air and ground vehicles, drones and robots.

[5]The further development of the process of specialization of the network of territorial and urban routes is characterized by new impulses in the recent proposals. Elon Musk's Project for a network of underground tunnels, bypassing the congested network of urban routes in the city of Los Angeles, characterized by mechanical tracks that transport high-speed vehicles is an interesting example (Cosimi 2017).

– integration of the network with new polarities dedicated to distribution and
multiple specializations (e.g.: commercial + residential + offices + hospitality
+ care).

By focusing more strictly on the urban scale, it is possible to look more closely
at the commercial suburban fabrics' transformation phenomenon.

Big portions of these fabrics are, on the one hand, no longer able to with-
stand, due to the lack of updating the competition of e-commerce, and on the
other hand, they can't compete with the traditional urban fabric store trade activities. While for a portion of them, considering a numerous and complex series of
factors (dimensions, catchment area, etc.), it is possible to hypothesize a transforma-
tive/conservative perspective to confirm the role of nodality/polarity even changing
its internal structure,[6] for the remaining fabrics more radical transformations, such
as total demolitions, or restructuring/transformation in distribution hubs could be
proposed.[7]

4.2 Aggregate Scale

The modern suburban shopping mall, as analyzed in Chap. 3 dedicated to the
formative process interpretation, reaches its mature form and type during the 1950s
and 1960s of twenty-first century. The *contrada* fabric structure that characterizes this aggregate reproduces in a more or less direct and clear way, the urbanity
of the traditional city integrated with some episodic, but strong, authorial critical
interpretations.[8]

The fabric of today's mall has a very low level of specialization; the route is
still the driving element in the hierarchical organization of this small piece of city.
Even when complete blocks exist, their rear/courtyard pertinent space is subdivided
between the units that face it and used for goods' storage and for technical functions,
similar to what observable in the basic block of traditional urban fabric, for example
in Rome (Caniggia and Maffei 1979, 136, tab. 24). Usually, these technical and
service functional rear spaces settle on a *technical pertinence route*, not accessible by
customers, but it may be that this route, allowing the transport of goods from unload
gate to the warehouses of each commercial units (without traveling the customers'

---

[6]Later, in the section dedicated to the transformation at the fabric scale, the modalities for this
transformation will be detailed.

[7]On transformation and reuse of large commercial surfaces, the so-called "big boxes": (Christensen
2008; Techentin 1991; Coleman 2006, 119).

[8]As regards the critical and authorial dimension the book refers to those experiences, many of
them chronologically located at the turn of 1970s and 1980s, characterized by an extremely free
interpretation of the type, as in the projects by the American architect Jon Jerde (1940–2015). Such
experiences would have paved the way for numerous proposals and built spaces by star system
architects, which until then had not been very attracted by the suburban retail topic, very influenced
by stringent rules of spatial organization.

actual phase

future phase

Fig. 4.1 Fabric's transformation scheme. *Source* V. Buongiorno

main commercial route), is located at an underground level and thus communicate vertically with the commercial units through elevator and stairs.

This base fabric scheme, characterized by the low level of *contrada* specialization, incorporates large surfaces for goods' storage, a "bay" for unloading, as well as individual smaller storage spaces for each commercial unit.

These existing storage spaces, with the changes and innovations in buying behavior produced by the advent of e-commerce, become quantitatively excessive, and constitute a resource, a precious surface reserved for the transformation of commercial suburban fabrics, for their necessary evolution, in a moment of strong crisis, like the current one.[9]

In the perspective of transformation, intended as a critical proposal for the continuation of the fabric's formative/transformative process, the commercial suburban

---

[9]The strong negative trend that characterizes mall activity; the large number of malls that close are the expression of a change in the way of life, purchasing and exchange, not followed in parallel by a typological update of the commercial/retail spaces (Teufel and Zimmermann 2015; Techentin 1991; Chung et al. 2001).

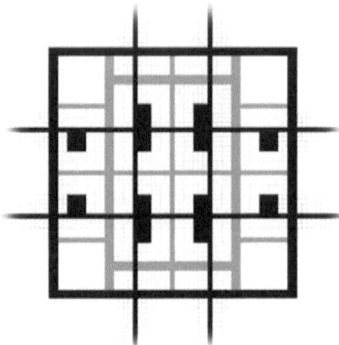

non-exclusively commercial routes and public spaces open during the **night**

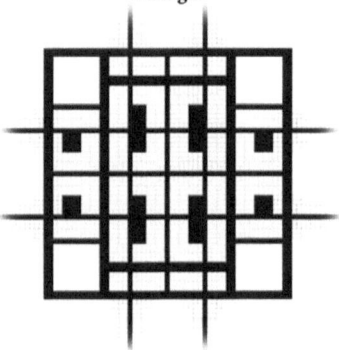

Commercial and non commercial routes and public spaces open during the **day**

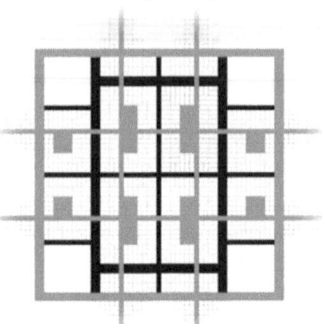

Commercial routes and public spaces open during the **day**

Fig. 4.2 Future phase fabric's routes. *Source* V. Buongiorno

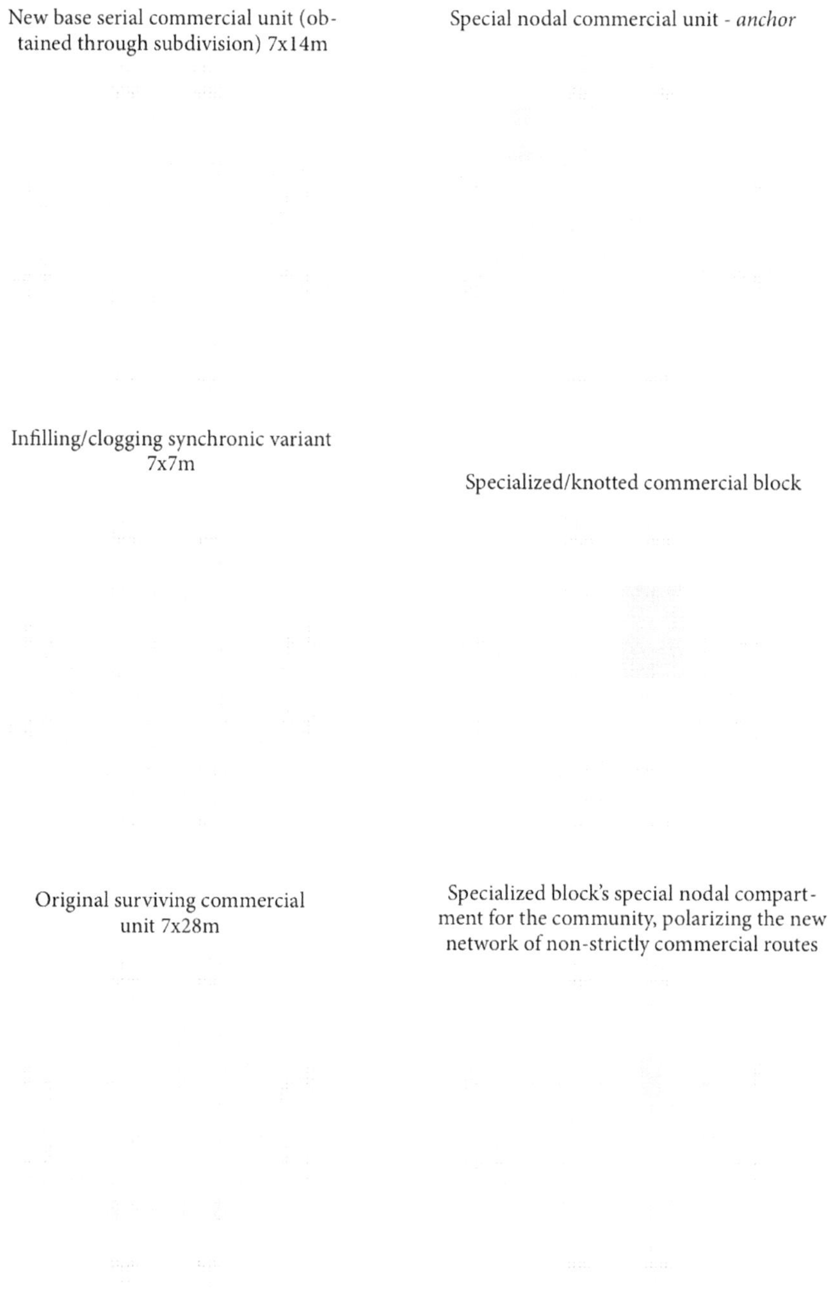

New base serial commercial unit (obtained through subdivision) 7x14m

Special nodal commercial unit - *anchor*

Infilling/clogging synchronic variant
7x7m

Specialized/knotted commercial block

Original surviving commercial
unit 7x28m

Specialized block's special nodal compartment for the community, polarizing the new
network of non-strictly commercial routes

Fig. 4.3 Future phase fabric's elements. *Source* V. Buongiorno

Commercial *contradas* - by day   Non-strictly commercial community *contra das* - night and day

Fig. 4.4 Future phase fabric's *contradas*. *Source* V. Buongiorno

fabric in crisis—since in its formative process it is possible to recognize the transformative dynamics and tools that traditional fabrics use to evolve and adapt organically to the new needs—could be a candidate to be object of a further transformation, through the specialization of the blocks dynamic typical of traditional urban fabrics, to increase the degree of internal hierarchical complexity (Figs. 4.1, 4.2, 4.3 and 4.4). This specialization, in analogy with what can be observed in traditional urban fabrics, can be implemented by overturning the commercial fabric routes inside the commercial blocks, by knotting them and consequently generating internal special nodal compartments that occupy the space previously occupied by the storage, technical spaces and related routes.

The new specialized blocks respond to the need for new spaces, which also take on the increasing demand for urbanity and community activities.

Once the specialization of the block is done, the material goods' purchase activity becomes less important, and the new commercial suburban fabric becomes:

– as regards small and medium-size commercial units, the specialized place for the sensorial and cultural experience of goods, experienced live first and then ordered/bought individually or directly from the store, to be delivered directly to customer's home (as was also in the case of the first department stores, like *la Rinascente* in Italy (Scarpellini 2008)). The mall becomes an increasingly less commercial place, more experiential and full of information about the multiple values of goods. It gives information about the productive process (industrial and artisanal), or about the lifestyle that each good and brand brings, in a way that has been already experimented and anticipated by the single-brand *concept stores* (Teufel and Zimmermann 2015);
– as regards special commercial units, a space in which the commercial aspect becomes even less important: these units are transformed and increasingly integrated into not exclusively commercial units, full of entertainment and leisure activities, or goods production and customization ones.

The non-commercial special units expand, thereby incorporating also special commercial units in transformation. The anchors of the new commercial special fabric are therefore made up of special mixed units in which the commercial activity is not prevalent. These new anchors include, instead, ample spaces for entertainment, food court and sport/physical activities.

The specialization of the blocks, at least some of them, leads to the formation of new special units, located in the special nodal rooms/compartment obtained through the base units recasting and fabric routes overturning inside the block (Fig. 4.3).

These new special units constitute new nodalities/polarities, whose strong vocation for community and public activities responds to the need for integration between these activities and the sphere of trade and consumption, including both the activities of physical exchange of goods and those of collective consumption such as entertainment or sport (contemporary examples of not exclusively commercial special Units, located in nodal spaces similar to those proposed, can be found in: *Mall of America*, Minneapolis and *American Dream mall*, Miami).

The new special rooms formed within the blocks constitute the poles/nodes of a new network of, not strictly commercial, routes (Figs. 4.2 and 4.4).

These routes network simultaneously: 1. It interconnects the special nodal rooms/compartments, located inside the specialized blocks; 2. It connects them horizontally with other routes located outside the commercial special fabric; 3. It connects them vertically, through the vertical overturning of horizontal routes, by elevators, ramps and escalators, with a possible extension of the fabric to upper floors (non-commercial and tending to basic building, with special housing, hospitality structures and offices).

While the formation of the special nodal compartment through the block's specialization is clearly comparable to the formative and transformative process of special buildings starting from the basic fabric (Palazzo, Monastery, Department store, etc.), the formation of the new network of horizontal and vertically overturned routes, instead, can be related to the process of formation of specialized routes or special fabrics, that integrate/overlap with the existing urban fabric and bypass it to carry out new functions that cannot be fully integrated with the existing fabric, as in the case of the Parisian *Passages* (Lemoine 1989; Geist 1995; Mackeith 1986; Benjamin 1986), underground North American parallel cities (Yoos and James 2016; Maitland 1985, 127–161) (e.g.: Montreal's *Ville Souterraine* (Boisvert 2011; Boivin 1989; Deglise 2008)) or at high altitude level (e.g.: *skywalks* Minneapolis).

But, if in Paris, Minneapolis and Montreal the parallel city discards or bypasses the city's major vehicular routes, too exposed to the harsh climate, too full of traffic and too noisy to provide comfort for pedestrians/buyers and also too wide to be easily used for a comparative shopping[10] walk, in the future commercial suburban fabric the new network of non-commercial routes instead bypasses the crowds of customers and shopping tourists. The new routes and the special nodal spaces that interconnect, even if accessible to all and integrated with the commercial routes, are particularly dedicated to those who reside on the upper floors (see: "Crystal city", Arlington, Virginia; Maitland 1985, 144) and to the surrounding urban fabric's inhabitants. The residential/mixed use urban fabric is the result of a process of reactivation/refilling (on the upper floors and in the adjacent areas it is likely to assume the establishment of offices, hospitality and special housing), which aims at a de-specialization of the commercial fabric conceived originally as strongly specialist.

Thus, the commercial suburban fabric is transformed, losing its strong suburban connotation, into a fabric with multiple and integrated specializations.

Nonetheless the specializations remain, a complete de-specialization would be, in fact, as difficult as it is useless for the purposes of transforming these fabrics from "selling machines" to "new urban nuclei" (Gruen 1973). A polarizing urban core for the sprawling residential suburb cannot, in fact, disregard the presence of special and basic functions integrated.

---

[10]Comparative shopping, a shopping activity which happens in a limited and comfortable space that can be traveled over in a limited time, allowing customers to make a choice from the various commercial units' goods offer, comparing them in a fairly short time.

There is in the outlined transformative horizon, the integration of other specializations (not only commercial), in coherence with the more general trend of contemporary retail, through the inclusion of community, leisure, entertainment, food and sport activities. The integration within the new whole fabric, in sections on the upper floors, of less specialized functions such as offices, hospitality and special housing, reflects the aim of integrating the special fabric as much as possible within a larger and more complex urban organism, going in the opposite direction from the modernist functional segregation.

The visitor just like the inhabitant of the new integrated multiple specializations fabric, therefore, once the transformation has been completed, reaches the new urban core, a new downtown for the periphery,[11] by using his private car or public railway transport, uses the intermodal exchange node, to switch on the pedestrian movement mode, by passing through the parking lot where the car is stationed and/or the railway station, finally accessing the fabric itself. The access to the fabric is no longer configured as access to a giant building through an undersized door, as happens in most cases of existing traditional mall. It is, instead, an urban gradual passage in which parts of fabric coexist with open air[12] paths and fully enclosed parts, almost as in a traditional urban fabric. The new fabric is distinguished by its previous version also because its internal routes extend outside, generating new urban fabric in the adjacent areas.[13] Probably not all the routes are prolonged toward the adjacent base fabric, but surely this is the case for non-commercial routes, polarized by non-commercial polarities located in the special nodal compartments resulting from blocks' specialization and knotting.

Therefore, by using the new routes, the future visitor/inhabitant, can access the multiple and integrated specializations fabric, coming from the adjacent base fabric by walking "routes" that are delimited by some commercial activities (clearly inferior in number and density and not comparable with those that characterize the main and secondary commercial routes), some services for residents, and punctuated by secondary nodes (the main ones are located in special nodal rooms) devoted to vertical overturning of the horizontal routes, useful to reach the non-commercial fabric at the

---

[11]On the construction of new centers for the peripheral city through the transformation of specialized fabrics and the experience of the American Edge cities see: Dunham-Jones and Williamson (2011, 172–200).

[12]As in the cases of Outlet villages (eg: McArthurGlen Designer Villages) and in the cases of malls which, also as a result of transformative interventions with demolition of parts of the roof, are characterized by a mixed situation, enclosed and open-air (eg: Eastgate mall, in Chattanooga, Tennessee-Usa); Coleman (2006, 112–121).

[13]An interesting example of this hypothesis is the Parco Leonardo Shopping Center, in Rome's hinterland. The plan of the center, drawn up by the Spanish architect Ricardo Bofill, foresees not only a shopping center but also an entire surrounding urban fabric, that complements the shopping center in an organic system. Despite an integration between the residential and commercial functions, that at the architectural/building scale was not perfectly successful, the project gave positive results. Located in a strategic position along the connection artery between the center of Rome, the international airport and the coastal settlements, the new fabric takes advantage of the railway connection and constitutes an attractor in commercial and residential terms for a vast catchment area.

upper floors. Continuing to walk horizontally they cross, in a free and unfiltered way, and/or with temporary closures during the night, the main and secondary commercial routes. Pre-existing to the transformation of the fabric, they maintain autonomy but are strictly connected to non-commercial routes. Services (toilets, etc.) are located in nodal positions, at the intersection (but not exactly in the corner space, which is too commercially valuable) between commercial and non-commercial routes, to be identifiable and easily accessible by any type of user.

After having crossed the commercial routes, the visitor heads toward the non-commercial nodality/polarity, located inside the new specialized blocks. These polarities are located in vast spaces, open air or covered with a transparent roof, delimited by a series of commercial and/or not commercial units, settled on the routes resulting from the pre-existing fabric's commercial routes overturning inside the block; the commercial units are both recast/passers-by between the original commercial route and the overturned new one, or independent, settled back-to-back. On the special nodal room/compartment in positions that are in turn nodal, near the intersections between routes, there are the service rooms (toilets and technical rooms) and the nodes of horizontal paths overturning in vertical ones that are polarized by the non-commercial fabric settled above on the upper floors. In the new special nodal compartments, several vertical and horizontal paths converge and interconnect them with the surrounding and above base fabric, urban and territorial organism.

From the special non-strictly commercial space, if the visitor was more interested in the commercial activity, he could walk to one of the commercial routes, passing by a commercial unit, and he would find a more or less wide and high path (depending on whether it is a main or secondary path), eventually surmounted by other levels that can be visually reached through openings in the floors. On these routes, both in the open air and in the enclosed stretches, there are base serial, intermediate and special commercial units. Commercial routes are polarized by special units, almost always positioned in nodal positions (nodalities generated by the intersection of main and secondary, commercial or non-commercial commercial routes). The commercial fabric's vertical growth (beyond the typical 3–4 floors) takes place especially in the node, as a place of fabric's horizontal routes overturning and vertical continuation, with the usual means of escalators or elevators, to allow integration, in section as in plan, of the non-commercial and commercial functions.

In the new transformed fabric, the vertical and tridimensional overturning of routes, far from reproducing in sections the segregation of pre-existing functions and kinds of users, generates a complex three-dimensional network that interconnect, both on ground floor and upper ones, commercial and non-commercial spaces.[14]

---

[14]The precedents for this fabric and paths' vertical overturning is not so similar to Chicago's Vertical Malls. It is more associable to specialized three-dimensional fabrics such as those in Hong Kong, Skywalks-Minneapolis and North American underground cities (Yoos and James 2016).

4.3 Building Scale

At the building scale of the transformed fabric, the new single base commercial units are apparently smaller, in comparison with those belonging to the pre-transformation fabric. Since they need a smaller space for goods' storage (due to the changes in the logistic and supply chain system), new units can count on a larger space that can be devoted to good's sensorial and cultural experience, bigger than that of the pre-existing fabric's units.

The same dynamic applies to intermediate aggregate and special units. The latter, located in polar/nodal positions within the fabric, become sort of hinges between strictly commercial, urban experiential and cultural/non-commercial aspects, celebrated above all in the special nodal rooms located inside the new specialized/knotted blocks. Some special units—anchors (as indeed already happened in the department store's formative process) occupy the entire specialized block, parts of it, or just its special nodal compartment/room.

If placed in an antipolar position at the fringes of the fabric, special units, accessible both from the outside and from the inside, also become hinges between the commercial fabric and the surrounding base fabric.

The internal micro-fabric of commercial and mixed-use units is the subject of a process of transformation and specialization, similar to what happens to the mall's urban fabric.

As regards the interiors, the commercial unit integrates the classic spaces for the display and for the goods sensorial experience (vertical elements for display, shelves and various other types) while other spaces focused on offering a cultural added value to the shopping experience. Thus, spaces for exhibitions and cultural events, for global information on products, for a coffee break or, as in the case of the *Prada 'epicenter'* store on New York's 5th Avenue (http://oma.eu/projects/prada-epi center-new-york—consulted on 02/08/2019), an auditorium for fashion shows and other cultural events, are organically inserted between the traditional stands for goods display.

This is a process of further specialization for commercial unit's spaces—already partially implemented in the case of special commercial units-anchors like department stores, with the coexistence within them of not strictly commercial complementary functions—but still not much present in serial base and intermediate aggregate commercial units.

If, as hypothesized, on the urban fabric scale, the critical process of further specialization can proceed with traditional fabric dynamics and methods, consequently it is possible to similarly think that, on the internal building fabric scale, the critical process could proceed to a transformation and specialization of the internal fabric by looking at its complete "blocks" formed by the display elements. On the other hand, even in the current situation, the serial, intermediate and aggregated commercial units concentrate the special goods, and therefore the "special" contents, in the "island"-type exhibitors. Thus, a specialization of the internal fabric's basic block can lead to the formation of a special internal nodal space, devoted to host complementary activity (e.g.: meetings and product presentations, rest/relax, etc.).

Last but not least, there is also the issue of the non-commercial polar/nodal parking lot's unit transformation. The spatial experience that is consumed within the commercial suburban fabric is largely influenced by the quality of parking spaces, by the distances to be traveled between the parking and the commercial fabric entrance, by the variety of spatial situations that characterizes the path to and from the parking to the entrance and that could be used to orientate customers in a context that is still too often made up of exclusive seriality. Thus, a transformation of the special parking unit can proceed through a specialization action, with the integration of other activities, more or less commercial (Underhill 2005, 31–32), within the parking spaces, in the perspective of hierarchizing a system that is still too serial and confusing.

4.4 Local/Global Retail Spaces Today—Reading and Transformation

The following paragraphs will be dedicated to the reading of two case studies, located in Rome and Quebec city. The two malls are read as special commercial suburban fabrics, in their actual shape the perspective of finding a transformational hypothesis.

The case studies are:

Quebec city, "Place Sainte Foy—Place de la Cité—Place Laurier" shopping centers complex;
Roma, "Galleria Porta di Roma" shopping center.

Both case studies are analyzed in their current phase. For both of them the analysis focuses on the aggregate scale, highlighting the routes, the fabric's contradas, and the nodalities. Finally, are considered blocks, with their transformation and specialization potential.

The following graphical interpretation is proposed as supporting and complementary material to the general theory outlined in the previous chapters.

4.4.1 "Galleria Porta di Roma" Shopping Center, Rome, Italy

This commercial complex has typical features of the regional mall, due to its isolated location in relation to the traditional urban fabric, near a specialized motorway artery (GRA—Grande Raccordo Anulare) which acts as a barrier for contact and communication of the with the urban fabric but also with the territory. Part of the wider urban project "Bufalotta", designed by Studio Valle Architetti Associati, is located exactly between the districts Val Melaina and Fidene-Castel Giubileo, next to the GRA (Figs. 4.5, 4.6, 4.7 and 4.8).

It is a mall of rather up-to-date design and this is demonstrated by the presence of special non-commercial units that diversify activities and related opening hours within the mall, and contribute to giving it the characteristics of a public space.

In its planimetric conformation, on the two floors in which it develops, it presents a main route that connects the special commercial and non-commercial units, multiplex cinema and technological goods' superstore, and other 2/3 orthogonal secondary routes that connect the mentioned special units with the large supermarket special unit-anchor. The routes are marked by nodal spaces with double and triple height in which the vertical routes (elevators, ramps, escalators) are located, to connect the two commercial floors and the two further parking floors below.

The commercial routes on their pertinent strips generate, following the typical behavior outlined in the previous chapters, a fabric that is characterized, in the inner part of the planimetric development, by complete blocks made up of serial/row commercial units, with the "courtyard" pertinent area space used as a technical space for technical and goods' storage functions. The fabric that can be observed shows a *"contrada"* organization: the block does not exist as a special, autonomous and identifying element within the system, every single serial commercial unit is identifiable as belonging to a route, instead of being part of a fabric block.

The internal covered routes are connected, in a relationship of continuity, to the external ones polarized by a further commercial anchor, formed by the Ikea superstore together with another large commercial unit. The articulation of these outdoor

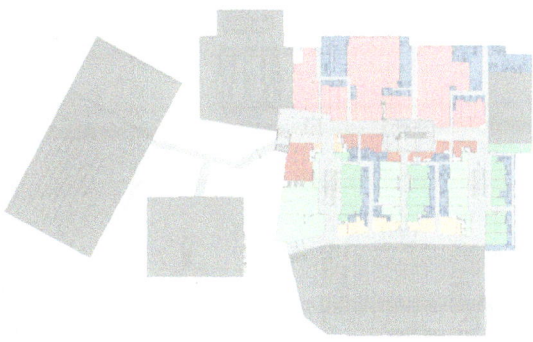

Fig. 4.5 Special commercial suburban fabric. Pertinent strips reading. *Source* V. Buongiorno

open-air itineraries includes an intermediate node where a special non-commercial entertainment unit (skating rink of small size but with a relevant role) is set up.

The portion of the building containing the covered commercial galleries acts as the basement for a tower of residences and offices on the upper floors. On the lower/underground floors, instead, there is the parking/intermodal change node.

The choice of this complex as a case study is due precisely to the updated and recent character of the center and to the fact that it presents the typical features of the regional mall (with the "contrada" organization and the typical low specialization of the blocks).

Moreover, the fact that the complex is located in an area close to the urban fabric but at the same time with a wide margin of expansion makes it interesting for the possibility of a transformation that foresees its expansion to the urban/territorial scale.

Fig. 4.6 Main commercial route (up), Secondary routes (down). *Source* V. Buongiorno

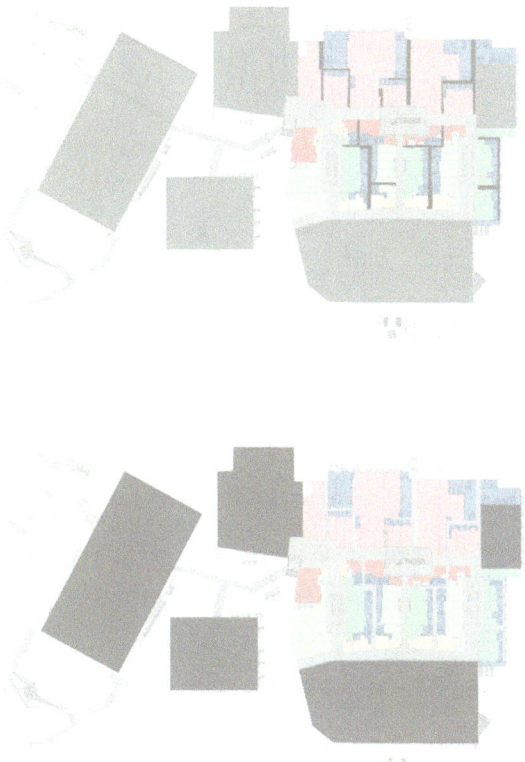

Fig. 4.7 Pertinent technical routes (up), Special nodal commercial units—*anchors* (down). *Source* V. Buongiorno

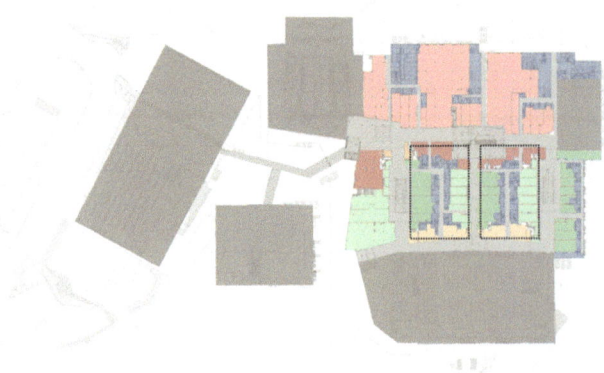

Fig. 4.8 Identification of fabric's complete blocks, for the further future transformation, specialization and knotting. *Source* V. Buongiorno

4.4.2 "Place Sainte Foy—Place de la Cité—Place Laurier" Shopping center, Quebec City, Canada

The complex of shopping centers is located on the Boulevard Laurier territorial route, next to the urban center of Sainte Foy. It is located at the meeting point of very important specialized routes, which passing through the Pont de Québec, on Saint Laurent river, and connect—by rail and automotive transport—Quebec city and Sainte Foy with several cities of the eastern North America, such as Montreal, Boston and New York. In the territorial nodality of Sainte Foy other functions coexist with commercial malls, such as that of urban and territorial scale equipments: the Universitè Laval campus, some important hospitals and office and tertiary buildings (Figs. 4.9, 4.10, 4.11 and 4.12).

Looking specifically at the complex of shopping centers, it is possible to see that it is characterized by a very articulated diachronic development (Moretti 1998, 1999; Lacroix 2003) that follows the general formative phases outlined in Chap. 3 (for a more detailed analysis see: Strappa and Buongiorno 2019).

In the current phase it is characterized by the organic set of three previously independent malls, organized on the same main commercial route.

Fig. 4.9 Special commercial suburban fabric. Pertinent strips reading. *Source* V. Buongiorno

As happens in most of current malls, and especially in situations less transformed
by the synthetic intervention of a single designer,[15] in the case of the Quebec City
Mall we can observe a *contrada* fabric with very few complete blocks.

The internal routes, including the main route shared by the three malls—continuous in the route track but discontinuous if we consider the caesuras generated by
alternating covered and open-air sections—are all disconnected from the surrounding
suburban fabric's routes network. Even though it is located at the center of the specialized fabric described above, the commercial fabric of "Place Sainte Foy—Place de la
Cité—Place Laurier" constitutes, as well as other contiguous special aggregates such
as the university campus, an island with limited accessibility (as regards accessibility
means and hours).

---

[15]In this sense it is possible to compare the Quebec city's case study, resulted from many mixed
interventions, not unitary and dictated by dominant market logics, to Rome's case studied, where the
spontaneous logic of commercial space coexists and is interpreted within the critical interpretation
of the unitary and synchronic plan designed by Studio Valle architects.

Fig. 4.10 Parking/intermodal exchange units (up), Special nodal commercial units—*anchors* (down). *Source* V. Buongiorno

Fig. 4.11 Main commercial route (up), Secondary routes (in the center), Pertinent technical routes (down) *Source* V. Buongiorno

in-depth study also in an urban context, so as to compose an organic frame, at every scale.

An analogous need for future integration of the research concerns the vertical/three-dimensional extension of special commercial fabrics. In this book, several references are made to the three-dimensional development of commercial fabrics, through the overturning of horizontal routes and the importance of such a dynamic of development in the formative process of special nodal commercial units and whole commercial fabrics (see also: Buongiorno (2019)).

Finally, the current extension of the range of complementary specializations combined with the purely commercial one in contemporary specialized fabrics suggests, for the future developments of the studies on this topic, the need to integrate the information related to these mixed specializations and to read—starting from the single specialized reading presented here—the most complex structure.

From the outlined frame of strengths and future potentialities of research emerges the fruitfulness and interest for investigating this topic area, and the evidence that further developments of the results proposed in this book and the extension of the field of investigation can contribute to enrich the endowment of reading and design disciplinary tools.

References

Buongiorno V (2019) Urban tridimensional special fabrics: formative and transformative process. In: ISUF 2019 XXVI international seminar on urban form: cities as assemblages proceedings. CyNUM, Nicosia Cyprus (publication in progress)

Strappa G (2017) Nuovi confini. Territori di ricerca della morfologia urbana. In: U + D Urbanform and design, n 07/08. L'Erma di Bretschneider, Roma

Fig. 4.11 Main commercial route (up), Secondary routes (in the center), Pertinent technical routes (down) *Source* V. Buongiorno

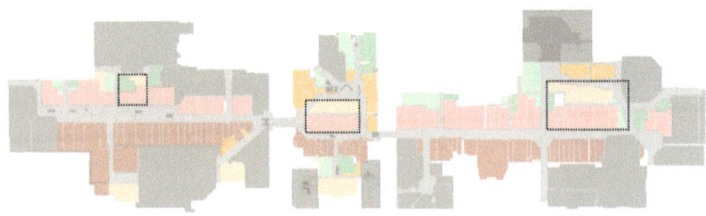

Fig. 4.12 Identification of fabric's complete blocks, for the further future transformation, specialization and knotting. *Source* V. Buongiorno

References

ATKearney agency's Report (2019) The future of shopping centers. (https://www.atkearney.com/documents/20152/986752/The+Future+of+Shopping+Centers.pdf/6455ae6f-f430-2fe7-285b-ef671153d29a-consulted on 21/04/2019)

Benjamin W (1986) Parigi, capitale del XIX secolo. I "Passages" di Parigi. Giulio Einaudi editore, Torino

Boisvert M (2011) Montréal et Toronto. Villes intérieures. La presse de l'Université de Montréal, Montréal

Boivin DJ (1989) Montréal souterrain: étude du réseau piétonnier protégé du centre-ville-Les Cahiers du CRAD, vol 12, n 1. Université Laval, Quebec-Canada

Caniggia G, Maffei GL (1979) Composizione architettonica e tipologia edilizia: 1. Lettura dell'edilizia di base. Marsilio, Venezia

Caniggia G, Maffei GL (1987) Composizione architettonica e tipologia edilizia: 2. Il progetto nell'edilizia di base. Marsilio, Venezia

Chung CJ, Inaba J, Koolhaas R, Leong ST (2001) The harvard design school guide to shopping/harvard design school project on the City 2. Taschen, Koln

Coleman P (2006) Shopping environnements. Evolution, planning and design. Elsevier Architectural Press, Oxford

Cosimi S (2017) In: LA, ecco il primo tunnel sotterraneo di Musk. https://www.repubblica.it/tecnologia/2017/10/30/news/elon_musk_ecco_il_primo_tunnel_sotterraneo_a_los_angeles-179759801/-consulted on 15/05/2019

Deglise F (2008) *Montréal souterrain. Sous le béton, le mythe.* Héliotrope, Montréal

Dunham-Jones E, Williamson J (2011) Retrofitting suburbia Urban design solutions for redesigning suburbs. Wiley, Hoboken-NJ

Geist JF (1995) *Le passage. Un type architectural du XIX siècle.* Pierre Mardaga éditeur, Ixelles-Belgio

Gruen V (1973) Centers for the urban environment: survival of the cities. Van Nostrand Reinhold, New York

Lacroix D (2003) Morphogenèse de Sainte-foy: le dilemme de la banlieue moderne entre structures héritées et forme urbaine nouvelle une analyse morphologique. Ulaval, Quebec

Lemoine B (1989) Les passages couverts en France. Délégation à l'Action Artistique de la Ville de Paris, Paris

Mackeith M (1986) *The history and conservation of shopping arcades.* Mansell publishing limited, London and New York

Maitland B (1985) Shopping malls: planning and design. Construction Press, London

Moretti G (1998) Analyse morphologique des centres commerciaux régionaux et des tissus urbains qui les contiennent: le cas de l'agglomération de Québec—Thèse de maîtrise. Université Laval, Québec city

Moretti G (1999) *Shopping centres and the urban fabric: the evolution of their relationships.* In: Corona R, Maffei GL (1999) Transformations of urban form: from interpretations to methodologies in practice. Proceedings of the Sixth International Seminar on Urban Form, Firenze, July 23–July 26-1999. Alinea editrice, Firenze

Scarpellini E (2008) L'Italia dei consumi. Dalla Belle Époque al Nuovo Millennio, Editori Laterza, Bari/Roma

Strappa G (2012) Studi sulla periferia est di Roma. Franco Angeli, Milano

Strappa G, Buongiorno V (2019) Commercial Urban Fabrics Updating. Retail globalization and Shopping cultural areas. In: Charalambous N, Zafer Cömert N, Hoşkara Ş (2019) CyNUM 2018 conference Urban Morphology in South-Eastern Mediterranean Cities: challenges and opportunities Proceedings. Cyprus Network of Urban morphology, Nicosia-CY

Techentin W (1991) Shopping mall: storia di un malessere. In: Lotus International, n 118. Editoriale Lotus, Milano

Teufel P, Zimmermann R (2015) Holistic retail design. Frame Publishers, Amsterdam, Reshaping shopping for the digital era

Underhill P (2005) Call of the Mall: the geography of shopping. Simon and Schuster, New York

White W (1991) L'idea della strada sensoriale. In: Lotus International, n 118. Editoriale Lotus, Milano Christensen 2008

Yoos J, James V (2016) Parallel cities: the multilevel metropolis. Walker art center, Minneapolis-MN

Chapter 5
Conclusions

The research proposed here, as mentioned in the introductory Chap. 1, has the goal of contributing to fill an important gap on the morphological investigation of the contemporary city and territory's fabrics (Strappa 2017, 2–5).

The work on the reading of the formative process and on the identification of scenarios in continuation of the process for future transformation and design actions is a partial contribution open to new elements that can complement the rather rich and complex framework.

The formative process interpretation/reading, even if open to enrichments and evolutions, constitutes a cognitive/scientific base, which can be used for the reading of special commercial fabrics located outside the western European–American cultural area taken into consideration in the book. It contributes to bridging the current scientific gap of instruments for the investigation on contemporary special built environments, which because of their relative "unfathomability" are simply entrusted to the category of the single and the fragmentary and passively accepted.

It is only through investigating the deep structure of such a widespread phenomenon that can become possible perspectives for a transformation or a logical, congruent and coherent project for new settlements. A project that becomes urgent, more than ever, within current crisis of traditional commerce. A crisis that, to become opportunity and future, must necessarily be a moment of reflection and research, on the problematic structures of the realities to be transformed. Only through investigation, the contemporary architect can understand the formative/transformative process and then propose his own, personal and unique, critical and creative reinterpretation of this process, the project.

The limited time and space of a short book have dictated the need to delimitate a potentially much broader research field.

Reading has been concentrated on the suburban regional and superregional shopping mall, instead of looking at its close urban correlation (e.g.: pedestrian mall). The morphological reading and its application on the case of special commercial fabrics have a global character and would require an extension of the discussion with an

© The Author(s), under exclusive license to Springer Nature Switzerland AG 2020
V. Buongiorno, *Suburban Retail Spaces*,
SpringerBriefs in Geography,
https://doi.org/10.1007/978-3-030-54991-6_5

in-depth study also in an urban context, so as to compose an organic frame, at every scale.

An analogous need for future integration of the research concerns the vertical/three-dimensional extension of special commercial fabrics. In this book, several references are made to the three-dimensional development of commercial fabrics, through the overturning of horizontal routes and the importance of such a dynamic of development in the formative process of special nodal commercial units and whole commercial fabrics (see also: Buongiorno (2019)).

Finally, the current extension of the range of complementary specializations combined with the purely commercial one in contemporary specialized fabrics suggests, for the future developments of the studies on this topic, the need to integrate the information related to these mixed specializations and to read—starting from the single specialized reading presented here—the most complex structure.

From the outlined frame of strengths and future potentialities of research emerges the fruitfulness and interest for investigating this topic area, and the evidence that further developments of the results proposed in this book and the extension of the field of investigation can contribute to enrich the endowment of reading and design disciplinary tools.

References

Buongiorno V (2019) Urban tridimensional special fabrics: formative and transformative process. In: ISUF 2019 XXVI international seminar on urban form: cities as assemblages proceedings. CyNUM, Nicosia-Cyprus (publication in progress)

Strappa G (2017) Nuovi confini. Territori di ricerca della morfologia urbana. In: U + D Urbanform and design, n 07/08. L'Erma di Bretschneider, Roma